Fatigue Life of Riveted Steel Bridges

Fatigue Life of Riveted Steel Bridges

Björn Åkesson

Consulting Engineer, Fagersta, Sweden

CRC Press
Taylor & Francis Group
Boca Raton London New York Leiden

CRC Press is an imprint of the
Taylor & Francis Group, an **informa** business

A BALKEMA BOOK

Cover photo: The old riveted railway bridge over the river Klarälven at Dejefors, Sweden. Span length: 80 m.

Taylor & Francis is an imprint of the Taylor & Francis Group, an informa business

© 2010 Taylor & Francis Group, London, UK

Typeset by Macmillan Publishing Solutions, Chennai, India
Printed and bound in Great Britain by Antony Rowe (A CPI Group Company), Chippenham, Wiltshire

All rights reserved. No part of this publication or the information contained herein may be reproduced, stored in a retrieval system, or transmitted in any form or by any means, electronic, mechanical, by photocopying, recording or otherwise, without written prior permission from the publishers.

Although all care is taken to ensure integrity and the quality of this publication and the information herein, no responsibility is assumed by the publishers nor the author for any damage to the property or persons as a result of operation or use of this publication and/or the information contained herein.

British Library Cataloguing in Publication Data
A catalogue record for this book is available from the British Library

Library of Congress Cataloging-in-Publication Data

Åkesson, B. (Björn)
 Fatigue life of riveted steel bridges / Björn Åkesson.
   p. cm.
 Originally presented as the author's thesis (doctoral)–Chalmers University of Technology, Göteborg, Sweden, 1994, under title: Fatigue life of riveted railway bridges.
 Includes bibliographical references and index.
 ISBN 978-0-415-87676-6 (hardcover : alk. paper) – ISBN 978-0-203-84776-3 (e-book)
 1. Railroad bridges. 2. Railroad bridges–Reliability. 3. Riveted joints–Testing. 4. Steel–Fatigue. I. Title.

TG445A37 2010
624.2′1–dc22
                                                                2010001385

Published by: CRC Press/Balkema
       P.O. Box 447, 2300 AK Leiden, The Netherlands
       e-mail: Pub.NL@taylorandfrancis.com
       www.crcpress.com – www.taylorandfrancis.co.uk – www.balkema.nl

ISBN: 978-0-415-87676-6 (Hbk)
ISBN: 978-0-203-84776-3 (eBook)

Contents

Preface and Acknowledgement   IX
Abstract   XI
List of symbols   XIII

1. Introduction   1
    1.1 Background   1
    1.2 Aim   3
    1.3 Scope and limitations   4
    1.4 Contents of this thesis   6

2. Previous Work – A Literature Survey   7
    2.1 Static behaviour of riveted connections   7
    2.2 Fatigue tests of riveted joints   7
    2.3 Full-scale fatigue tests of riveted bridge members and joints   8
         2.3.1 "Experiments to determine the effect on impact, vibratory action, and long-continued changes of load on wrought-iron girders" (Fairbairn 1864)   9
         2.3.2 "Fatigue life extension of riveted connections" (Reemsnyder 1975)   10
         2.3.3 "Comportement à la fatigue de profiles laminés avec semelles de renfort rivetées" (Rabemanantsoa and Hirt 1984)   12
         2.3.4 "Fatigue strength of weathered and deteriorated riveted members" (Out, Fisher and Yen 1984)   13
         2.3.5 "Fatigue of riveted connections" (Baker and Kulak 1985)   14
         2.3.6 "Fatigue and fracture evaluation for rating riveted bridges" (Fisher, Yen, Wang and Mann 1987)   15
         2.3.7 "Fatigue strength of corroded steel plates from old railway bridge" (Abe 1989)   17
         2.3.8 "Fatigue and fracture of riveted bridge members" (Brühwiler, Smith and Hirt 1990)   17

		2.3.9	"Experimental and theoretical investigations of existing railway bridges" (Mang and Bucak 1990)	18

        2.3.10 A comparison and summary of the results from the full-scale fatigue tests   19

3. Riveted Connections   23
   3.1 Historical review   23
   3.2 Riveting technique   24
   3.3 Load-carrying capacity   25
       3.3.1 Clamping force   25
       3.3.2 Mechanical properties   30
       3.3.3 Shear strength   32
       3.3.4 Tensile strength   35
   3.4 Fatigue life of riveted connections   38

4. Fatigue of Riveted Railway Bridges   45
   4.1 General   45
   4.2 Fatigue critical details   48
   4.3 Dynamic amplification   56
   4.4 Brittle fracture   58
   4.5 Corrosion   68
   4.6 Inspection and maintenance   70
   4.7 Reinforcement methods   71

5. Field Studies   73
   5.1 General   73
   5.2 Visual inspection   75
   5.3 Strain measurements   75
   5.4 Deflection   77
   5.5 Dynamic amplification   79
   5.6 Crack control   83
   5.7 Steel samples   86
   5.8 Material testing   87
   5.9 Load-carrying capacity   91
       5.9.1 General   91
       5.9.2 Design load   91
       5.9.3 Remaining fatigue life   92
       5.9.4 Fracture mechanics analysis   99

6. Full-scale Fatigue Tests of Riveted Girders   103
   6.1 General   103
   6.2 Test specimens   103
   6.3 Test set-up and testing procedure   108

	6.4	Test results		113
		6.4.1	Fatigue loading	113
		6.4.2	Tension tests	123
		6.4.3	Impact testing	126
		6.4.4	Chemical analysis	127
		6.4.5	Clamping force	127
		6.4.6	Fatigue damage accumulation	130
	6.5	Comparison with other investigations		134
	6.6	Future investigations		135
7.	Summary and Conclusions			137
8.	Modern Research			143

Examples   147
Literature   155
Index   165

Preface and Acknowledgement

This book was originally published in 1994 by the Department of Structural Engineering, Division of Steel and Timber Structures, at Chalmers University of Technology in Göteborg, Sweden (with the title "Fatigue Life of Riveted Railway Bridges"). It was my doctoral thesis, the result of six years research. In order to bring it up to date, the findings of research conducted in this field in between 1995–2009 have been added (both in a separate chapter – Chapter 8 Modern Research – and as additional references) together with a set of examples, a list of symbols and an index. In fact, the set of examples summarize the entire thesis. In collecting material on the research done in this area in recent years, I have had immense help from associate professor Mohammad Al-Emrani at Chalmers University of Technology – what would I do without your help Mohammad! There were a lot of people that I thanked in the preface of the 1994 version, and I must once again repeat my deepest gratitude to my supervisor back then, Professor Bo Edlund (today professor emeritus), for his untiring support and total commitment, but this time I will take the opportunity to acknowledge those that mean the most to me, namely my three children, Josefin, Matilda and Petter. At the public defence of my doctoral thesis, when Josefin was just little more than 18 months old, she broke free from her mother (who had to concentrate on our second child, our newly born daughter Matilda) and she came up to me on the podium just after the presentation had started, which in fact was quite a nice interruption as it stopped the ceremony for a while and made me relax. This very early interest in her father's research showed itself when three years later, at the age of four, she asked me this question when we were passing under a bridge: "How do you build up there, dad?" Quite a relevant question, and I then did my best to explain how a bridge is built. In fact, later on I used Josefin's question for many years at the start of my introductory lecture in bridge engineering for the first-year students at Chalmers. When it comes to Matilda, she stunned her father at a very early age with reflections on life – reflections that he never would have come up with himself. At the age of eleven she started to teach her father the possibilities of the computer program Microsoft Paint. Whenever I said that something was impossible to draw she just said: "Let me show you dad!" Finally, when it comes to my son Petter (today ten years old), he has for many years built innovative and complex structures in Lego, which really impresses his structural engineer of a father – structures that his father was incapable of producing at the same age (being a Lego builder himself as a boy). So, at the same time as I was teaching master's students coming from all over the world, and later on when I started writing books, I was learning and getting inspiration from my own children, and I still am.

January 2010
Björn Åkesson

Abstract

This dissertation, which focuses on the fatigue life of riveted steel railway bridges, is based on three main elements:

- A series of full-scale fatigue tests of nine riveted stringers taken from a railway bridge built in 1896. A review of, and comparison with, fatigue tests on riveted railway bridge members that have been conducted in other laboratories over the years is also given.
- A number of field tests on riveted railway bridges. Some 15 bridges have been investigated by the author since 1989.
- A theoretical study of the load-carrying capacity of riveted connections. Special emphasis is given to the influence of the clamping force. The phenomenon of fatigue cracking in riveted railway bridges is also accounted for. The structural members and locations where fatigue cracking is likely to occur are a particular focus of this study.

The main finding is that there seems to be substantial remaining fatigue life in the riveted railway bridges still in use today. The full-scale fatigue tests have shown that the common standard fatigue design curve for riveted members and details (virgin specimens) is underestimating the fatigue life. The fatigue damage accumulation of the old bridge stringers, as measured in the laboratory tests, was found to be negligible, i.e. the stresses due to the in-service loading history have in general been low. In situ strain measurements on several riveted railway bridges also show that the stresses seldom, if ever, exceed the fatigue limit for riveted details. Ultrasonic testing, which has shown an absence of fatigue cracks, confirms these conclusions.

The fracture toughness is in general low for the steel in old bridges, but there are a number of circumstances that indicate that the ductility is adequate despite this fact. For example, low in-service stresses, low strain rates, the absence of fatigue cracks or other major defects, small plate thicknesses and an inherent structural redundancy of built-up riveted bridge members all contribute to a safe behaviour with respect to brittle fracture. In addition, the inherent structural redundancy of the riveted members has, during the full-scale fatigue tests, led to the arrest of a propagating fatigue crack when passing from one member part to another.

One main conclusion is that old riveted railway bridges do not have to be replaced as soon as is often the case today. The assumed presence of fatigue cracks and the fear of brittle fracture are often unjustified.

Keywords: *Railway bridges, steel bridges, fatigue life, full-scale fatigue tests, field tests, riveted connections*

List of symbols

α	angle
γ_{Ff}	partial safety factor (fatigue loads)
γ_{Mf}	partial safety factor (fatigue strength)
$\Delta\sigma$	cyclic stress [MPa]
$\Delta\sigma_D$	constant amplitude fatigue limit [MPa]
$\Delta\sigma_L$	cut-off limit [MPa]
Δt	time interval [s]
ε	strain [%]
θ	rivet diameter [mm]
μ	friction coefficient
σ	stress [MPa]
σ_B	tensile strength [MPa]
σ_{cl}	clamping stress [MPa]
σ_i	initial tension [MPa]
σ_r	stress range [MPa]
σ_{ru}	fatigue limit [MPa]
σ_y	yield stress [MPa]
A	elongation [%]
A	cross-section area [mm²]
A	protruding rivet shank length [mm]
a	crack length [mm]
a_c	critical crack length [mm]
b	width [mm]
C	category detail factor
C	constant
c	rivet spacing [mm]
D	category detail factor
D	fatigue damage accumulation factor
D	diameter [mm]
d	diameter [mm]
E	modulus of elasticity [MPa]
F	axial force [kN]
F	tensile force [kN]
F_{clamp}	clamping force [kN]

xiv List of symbols

F_{ext}	external load [kN]
F_{prying}	prying force [kN]
F_r	loading range (cyclic load) [kN]
F_{sep}	separation load [kN]
f_6	nondimensional factor
g	grip length [mm]
I	second moment of area [mm^4]
I	influence line coefficient
K	stress concentration factor
K_c	fracture toughness [MPa$\sqrt{m}$]
K_I	stress intensity factor [MPa$\sqrt{m}$]
ΔK	stress intensity variation [MPa$\sqrt{m}$]
k	quotient
L	length [mm]
L	grip length [mm]
ΔL	resulting elongation [mm]
ΔL_{plate}	compression of connected plates [mm]
ΔL_{rivet}	residual rivet elongation [mm]
ΔL_{temp}	thermal contraction [mm]
M	moment [kNm]
M_r	moment range [kNm]
N	number of loading cycles until failure
N	normal force [kN]
N_r	normal force range [kN]
n	number of loading cycles
n	number of slip planes
n	number of rivets
P	joint load [kN]
P	point load [kN]
P_{max}	maximum load [kN]
P_r	load range [kN]
P_u	ultimate load [kN]
P_{ult}	ultimate load [kN]
P_{slip}	slip load [kN]
p	slip deformation [mm]
R	stress ratio
R	reaction force [kN]
R_{eL}	yield point [MPa]
R_m	tensile strength [MPa]
r	radius [mm]
S	first moment of area [mm^3]
S_{plate}	plate stiffness [kN/m]
S_{rivet}	axial stiffness of rivet [kN/m]
t	thickness [mm]
t	temperature [°C]
V	shear force [kN]

Chapter 1

Introduction

1.1 Background

At the beginning of the 1990s (when the thesis was written), the Swedish National Rail Administration managed around 1100 steel railway bridges. Most of these railway bridges (approximately 800) were built in the period 1895–1940 (Fig. 1.1).

The majority of the railway bridges built in Sweden before 1940 were of riveted construction (say about 500–600 bridges) and most of these riveted railway bridges still remained in use at the beginning of the 1990s even though their average age was considerable and also despite the fact that they were being exposed to loading conditions (i.e. higher axle weight) greater than originally intended. The main reasons why it was possible to upgrade the loading capacity of the riveted railway bridges after they had been in service for many years were:

- At the beginning of the twentieth century there were frequent changes in the axle weight of locomotives and trains, therefore the bridges had to be designed with a certain "over-capacity" in order to meet this anticipated increase in loading condition.
- There had also been a successive increase in allowable stresses in the structural codes.

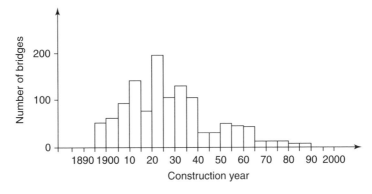

Fig. 1.1 Histogram of steel railway bridges in Sweden with respect to their construction year (data from the beginning of the 1990s).

- The prescribed load combinations for the design of a railway bridge became more and more realistic, and based on probability rather than "maximum combined effect"
- The dynamic amplification factor in the codes had varied over the years, but by the early 1990s was lower than those specified in the codes valid for earlier periods (Fig. 1.2).

The average age of the riveted railway bridges which still remained in use in the early 1990s was about 70 years and there were reasons to presume that their useful lifespan was close to its end. In the early 1990s, it was common practice to replace the entire bridge if it had been judged as "insufficient", while in older days repairs or strengthening work were undertaken in order to keep the bridge in service. These are the main reasons behind the decision to replace an old riveted railway bridge:

- Insufficient load-carrying capacity.
- The need to arrange ballast under the track on the superstructure.
- A fear of fatigue cracking.
- Excessive deformations when loaded by passing trains.
- The presence of defects such as loose rivets, corrosion or "hit marks".
- A fear of brittle fracture together with a lack of knowledge concerning the effect of ageing (i.e. the embrittlement of steel).
- High service and maintenance costs.

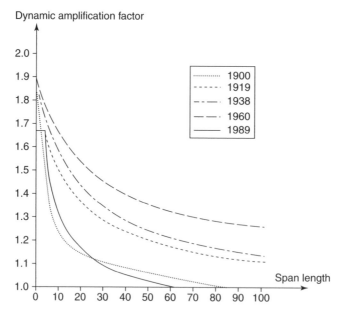

Fig. 1.2 The dynamic amplification factor as a function of span length in the Swedish codes for railway bridges. By comparing the different curves one can see the gradual change in the codes throughout the years.

Research concerning load-carrying capacity and lifespan expectancy of riveted and welded railway bridges were conducted at the Department of Structural Engineering, Division of Steel and Timber Structures at Chalmers University of Technology (see refs. [1]–[10]) in the late 1980s and early 1990s. Several field investigations of riveted railway bridges were made by the author (see refs. [1], [3], [5] and [9]), with 15 different bridges studied during a five-year period between 1989 and 1993. The results from these different investigations can be summarized in these principal findings:

- All the bridges were in a fairly good condition – even if the amount of corrosion differed substantially between the bridges.
- The bridges have been able to carry greater loads than the original design load.
- The stresses derived from the strain measurements were in most cases lower than the assumed fatigue limit for riveted details (i.e. when the fatigue life is infinite).
- The dynamic amplification of the measured strains due to passing trains was rather low when compared to calculated static values.
- During the fatigue crack control of the riveted connections, which was conducted using ultrasonic testing equipment, no cracks were detected.
- The fatigue damage accumulation analysis showed, in general, a substantial remaining fatigue life.

Field investigations of existing riveted railway bridges as well as full-scale fatigue testing of riveted members from railway bridges taken out of service engaged several researchers throughout the world in the twenty years prior to these 1990 studies. Among others, one can especially mention John W. Fisher at Lehigh University, Bethlehem, USA and Manfred A. Hirt at EPFL, Lausanne, Switzerland. These two contributed immensely to the knowledge concerning these issues and also inspired other researchers in the field.

1.2 Aim

Many old riveted railway bridges in the early 1990s were replaced too soon due to a general lack of knowledge about their expected lifespan. This indicated a need for more research concerning fatigue and brittle fracture in these bridges. The bridges had in many cases been judged unfit with reference to fatigue – although no analysis of the fatigue damage accumulation was made, and what is even more surprising, the assumed presence of fatigue cracks was not confirmed. In addition, no investigation was made into the possibility of strengthening the structure in order to increase its load-carrying capacity and, at the same time, to improve its resistance against fatigue. One main intention of this dissertation was to show how it is possible to take fatigue properly into account when deciding about the future service life of a riveted railway bridge.

The aim of the dissertation was to show that there was a substantial remaining fatigue life in the riveted railway bridges still in use in the early 1990s (focusing mainly on the primary stresses). Questions that will be discussed and answered include:

- What is the expected fatigue life of a riveted railway bridge member?
- What is the level of stress ranges expected for everyday train traffic?

- Is it possible to assume a very long (almost infinite) fatigue life for different riveted members subjected to stress ranges below the fatigue limit?
- How does temperature and the effect of ageing influence fatigue crack propagation?
- How should one take different loading histories into consideration when carrying out a fatigue damage accumulation analysis?
- How should a proper fatigue crack detection control be made?
- In what parts and at what locations in the structure are fatigue cracks to be expected?
- After how many stress cycles is a fatigue crack expected to become visible?
- How are loose rivets affecting the fatigue life of different bridge members?
- How should the probability of defects other than fatigue cracks be taken into consideration?
- Is it possible to repair and strengthen these old riveted structures?
- What should a proper instruction manual for the inspection routines concerning fatigue cracks look like?

Through extensive laboratory fatigue tests of riveted stringers from a railway bridge (see section 1.3 "Scope and limitations" and Chapter 6 Full-scale Fatigue Tests of Riveted Railway Girders) it should be possible to answer all of these questions.

How was it possible to justify research on fatigue and lifespan expectancy of riveted railway bridges? The railway bridges were soon taken out of service anyway, and in addition to that, the phenomenon of fatigue in structures subjected to variable loading in time had been well known to researchers for a long period of time (for 50 years or so, relative to the early 1990s). It was true that a great amount of research had been done, but it mainly focused on bolted and welded structures and their performance. Very little attention had been given to riveted details and especially little to riveted full-scale structures. In the last decades (before the early 1990s), riveted railway bridges had often been left without proper maintenance and very little funding had been made available to research programmes concerning these kinds of structures. One could say that the research in the period up to the early 1990s had been focused on the development of new materials and structures rather than the further development of existing structures.

Research concerning fatigue and lifespan expectancy of old existing riveted railway bridges in steel should enable the railway authorities to plan and decide exactly when and which of these bridges should be replaced or strengthened. In times of environmental awareness more and more attention is paid to the waste of resources. Prolonging the use of the old stock of riveted railway bridges by the help of this kind of research should not only give the railway authorities, but also society at large, considerable economic savings.

1.3 Scope and limitations

This thesis mainly concerns old riveted railway bridges with regard to fatigue and lifespan expectancy. There are of course many other railway bridge structures (e.g. bridges with welded or rolled girders, joined by welding or bolting) where fatigue is of great concern, but for riveted connections there are some special and important parameters where there is great uncertainty about their quality characteristics, such as:

- The clamping force in riveted connections is of lower magnitude and varies more than the same property in high-strength bolted connections.
- A riveted and built-up I-beam consists of different parts joined together in order to achieve composite action. In comparison to welded or rolled I-beams, riveted built-up I-beams possess an inherent structural redundancy when subjected to a propagating fatigue crack.
- Without exception, riveted construction was the rule for sizeable steel railway bridges at the beginning of the twentieth century and in the first few decades thereafter. The age of these bridges indicates that the steel type and steel quality is uncertain and, consequently, that the strength of the bridge members is of a questionable magnitude, particularly in comparison to more modern steel structures that have members with more defined and controllable quality properties.
- From the fact that riveted railway bridges in general are rather old and have been in service for a long period of time one can also assume the presence of corrosion damage, which can affect the load-carrying capacity and fatigue resistance. With respect to brittle fracture, one must also take the possibility of ageing into account.

The specific riveted members chosen for the laboratory investigation into fatigue and lifespan expectancy (reported in Chapter 6) were 36 built-up I-beam stringers taken from an old railway truss bridge built in 1896. There are, of course, several other components in a truss bridge that are subjected to fatigue due to variable amplitude loading (e.g. the truss bars and the floor beams). The reason behind the choice of the stringers as test objects was the simple fact that – due to their shorter influence length (in comparison to truss bars and floor beams) – these members have, over the years, been subjected to the largest amount of fatigue damage accumulation in this type of bridge. This fact indicates that it is often the stringers that govern the fatigue life of a railway bridge, and they are therefore the most important bridge member to study.

The results from the laboratory investigation into fatigue and lifespan expectancy of stringers taken from a riveted railway truss bridge are not only applicable to this kind of structural member. There is an applicability of the results to every open-deck construction (i.e. where the sleepers are directly placed on the main carrying bridge members), such as a simply supported or continuous riveted girder bridge (railway or highway bridge) of similar design.

There are also, of course, questions concerning the applicability of these results for other bridge types with differently designed bridge members. In order to concentrate the research efforts, we had to take all the stringers from the same bridge. Instead of taking members from a number of different bridges, we could by this procedure gain two advantages:

- All stringers had exactly the same loading history.
- There were reasons to believe that the stringers from this bridge had the same steel material characteristics.

Are the results from these full-scale fatigue tests only applicable to riveted railway bridges in Sweden, given the environmental conditions in the country? Of course, the types of construction and the properties of steel used in bridge building can vary from country to country, but these variations are also found between different bridges of the same "nationality". The problems concerning fatigue and lifespan expectancy of riveted railway bridges are of a universal applicability.

The connections between the longitudinal stringers and the transverse floor beams were outside the scope of this investigation. They are, however, very important and critical details for this type of bridge are discussed in section 4.2, where results of more recent research are considered.

1.4 Contents of this thesis

Chapter 1 gives the background and also the relevant motivation and need for this research. In Chapter 2 there is a summary of previous full-scale fatigue tests on riveted bridge members that have been conducted over the years at other laboratories. The load-carrying behaviour of riveted connections under static loading is analyzed and discussed in Chapter 3. Special emphasis is given to the clamping force. In Chapter 4 the phenomenon of fatigue cracking in riveted railway bridges is accounted for. The results from different field studies on riveted railway bridges that have been performed by the author are reported in Chapter 5. In Chapter 6 a full-scale fatigue test series of nine riveted railway bridge stringers is presented. The results and the conclusions that can be drawn from these results are summarized in Chapter 7. Finally, in Chapter 8 there is a compilation of full-scale fatigue tests on riveted bridge members carried out in the period 1995–2009.

For a reader who wants a quick summary of the results, there is an answer to all the questions posed in section 1.2 ("Aim") at the end of Chapter 7.

Chapter 2

Previous Work – A Literature Survey

2.1 Static behaviour of riveted connections

Static tests as well as theoretical studies of riveted connections have, since the riveting technique came into use in steel construction, been extensively performed. The investigations have mainly focused on the tensile and shear strength of riveted connections and also on the load distribution among fasteners. Factors such as bearing ratio, clamping force, slip resistance and rivet pattern have been taken into consideration.

When investigating the load distribution among fasteners in a riveted connection, it has been found that the end fasteners carry a larger proportion of the load than the interior ones, and this is due to a difference in elongation of the parts of the plate between the rivets.

The clamping force from the rivets and the ability of the joint to resist slip were found to depend upon factors such as grip length, the coating on the interfacing plates and the rivet installation procedure.

2.2 Fatigue tests of riveted joints

Numerous laboratory tests have been performed over the years to determine the fatigue strength of riveted connections (see e.g. refs. [59], [101] and [102]). Important parameters such as the residual clamping force and the bearing ratio were shown to greatly influence the fatigue resistance.

In order to obtain fatigue data on riveted joints and connections, the general idea has been to conduct tests on smaller details rather than full-scale bridge members. The reasons for this are easy understandable:

- Smaller details are easier to reproduce in a great number. This makes it possible to ensure a proper statistical evaluation.
- When performing fatigue tests on small riveted details, it is possible to control and alter the parameters that influence the fatigue strength.
- Smaller specimens are also easier to handle in the laboratory.

2.3 Full-scale fatigue tests of riveted bridge members and joints

Due to the reasons mentioned in section 2.2, full-scale fatigue tests of riveted bridge members and joints are scarce and the results from these tests have, to the author's knowledge, not been used as a basis for fatigue design curves.

The necessity of conducting a full-scale fatigue test on a real riveted bridge member or joint is two-fold:

- By testing a built-up riveted member you take into account the inherent structural complexity which arises from the composite action of the different member parts.
- The result will show the "overall behaviour", i.e. the interaction of different influencing parameters.

In order to attempt a qualified and correct estimation of the fatigue life of a railway bridge, investigators should take into account actual behaviour rather than utilize the results from "isolated parametric studies" on small, specially fabricated specimens.

Only a few number of full-scale fatigue tests on riveted bridge members, reported in conference papers or in scientific journals, have been found by the author, namely:

- Fairbairn, W.: "Experiments to determine the effect on impact, vibratory action, and long-continued changes of load on wrought-iron girders", 1864 [17].
- Reemsnyder, H.S.: "Fatigue life extension of riveted connections", 1975 [18].
- Rabemanantsoa, H. – Hirt, M.A.: "Comportement à la fatigue de profiles laminés avec semelles de renfort rivetées", 1984 [20].
- Out, J.M.M. – Fisher, J.W. – Yen, B.: "Fatigue strength of weathered and deteriorated riveted members", 1984 [21].
- Baker, K.A. – Kulak, G.L.: "Fatigue of riveted connections", 1985 [22], cf. also [23].
- Fisher, J.W. – Yen, B.T. – Wang, D. – Mann, J.E.: "Fatigue and fracture evaluation for rating riveted bridges", 1987 [24], cf. also [26].
- Abe, H.: "Fatigue strength of corroded steel plates from old railway bridge", 1989 [25].
- Brühwiler, E. – Smith, I.F.C. – Hirt, M.A.: "Fatigue and fracture of riveted bridge members", 1990 [27], cf. also [28].
- Mang, F. – Bucak, Ö.: "Experimental and theoretical investigations of existing railway bridges", 1990 [29], cf. also [30–32].

These different studies are summarized in this chapter, and the results from these tests are compared with the fatigue design specifications (i.e. fatigue design curves emanating from "small" laboratory tests).

Of the nine investigations reviewed here, only the five presented in sections 2.3.4 and 2.3.6–2.3.9 are full-scale tests of riveted girders taken from old bridges in service. Two investigations, sections 2.3.2 and 2.3.5, report tests on old riveted truss members. The remaining two investigations, sections 2.3.1 and 2.3.3, were on riveted girders that have not been in use. The different stress ranges given in this chapter are assumed by the writer to be at the outermost edge of the girders (and not at the actual rivet locations).

Table 2.1 The results from the first series of fatigue tests conducted by Fairbairn in 1864 [17].

Step	Loading P_{max}	Equivalent stress range σ_r	Cycles n_i	Accumulation Σn_i
1	$1/4 \cdot P_u$	56 MPa	596 790	596 790
2	$2/7 \cdot P_u$	63 MPa	403 210	1 000 000
3	$2/5 \cdot P_u$	89 MPa	5 175	1 005 175

2.3.1 "Experiments to determine the effect on impact, vibratory action, and long-continued changes of load on wrought-iron girders" (Fairbairn 1864) [17]

As early as 1864 Fairbairn presented the results of a test on a riveted wrought-iron built-up I-beam [17]. The purpose of this investigation was to study long-term effects on the "breaking" strength during cyclic loading. The phenomenon of fatigue was at that time not fully understood, but in general the view was that the static strength of a material was reduced in some way when subjected to repeated loading. This was explained by "its loss of cohesive powers", i.e. the material was supposed to become more and more brittle during cyclic loading:

> "It has been assumed, probably not without reason, that wrought iron of the best and toughest quality assumes a crystalline structure when subjected to long and continuous vibration – that its cohesive powers are much deteriorated, and it becomes brittle, and liable to break with a force considerably less than that to which it had been previously subjected." [17].

In order to study this effect, and at the same time determine an acceptable level of cyclic loading with respect to the risk of fatigue, Fairbairn tested a specially fabricated 6.1 m long and 410 mm deep riveted built-up I-beam. The beam had a 3 mm thick web plate and each flange consisted of two L-shaped profiles (L $50 \times 50 \times 5$ mm^3) with an additional flange plate (100×6 mm^2) attached to the L-profiles. He let the simply supported beam be subjected to a repeating load at its midspan and then he gradually raised the level of this cyclic load in order to find the "cyclic breaking strength". Fairbairn found that when he raised the load to two fifths of the ultimate static breaking strength, the number of load cycles until complete failure of the tension flange was sufficiently low to determine that this level of cyclic loading was unacceptable with reference to providing safety against fatigue. Fairbairn pointed out that it was the steel plate and the profiles that broke, and not the rivets. This fact was satisfying according to him. The results of this first series of tests are shown in Table 2.1.

In order to continue the testing, he had the beam repaired by replacing the broken tension flange and by adding an extra reinforcing plate on top pf the plate where the web had cracked. In this second series of tests he started at the same loading level as the previous test, but then he had to raise the level to one third of the ultimate static breaking strength in order to obtain a fatigue failure (Table 2.2).

Table 2.2 The results from the *second* series of fatigue tests conducted by Fairbairn in 1864 [17].

Step	Loading P_{max}	Equivalent stress range σ_r	Cycles n_i	Accumulation Σn_i
1'	$1/4 \cdot P_u$	56 MPa	3 150 000	3 150 000
2'	$1/3 \cdot P_u$	74 MPa	313 000	3 463 000

Fairbairn summarized his results with these remarks:

- Wrought-iron beams have insufficient strength against fatigue if they are loaded at a level of one third or more of the static breaking strength (i.e. $\sigma_r \geq 74$ MPa).
- Identical beams have sufficient strength against fatigue if they are loaded at a level of a quarter or less of the static breaking strength (i.e. $\sigma_r \leq 56$ MPa).
- The highest allowable stress level of static loading recommended by the railway authorities, "five tons per square inch" (76 MPa), can be seen as a reasonable level with reference to ensuring safety against fracture.

2.3.2 "Fatigue life extension of riveted connections" (Reemsnyder 1975) [18]

From 1970 onwards, the number of full-scale fatigue tests of riveted bridge members increased, although the numbers remained small. One of the first investigators during this period, and the one who carried out the most extensive tests, was Reemsnyder, who in 1975 [18] conducted a series of fatigue tests of 18 riveted truss members (16 specially fabricated full-scale specimens and two actual bridge members taken from service from an ore bridge built in 1917). The aim was to study the effect of replacing the rivets in critical regions with high-strength bolts after fatigue cracking. Reemsnyder also studied the effect of:

- Different constant-amplitude load levels.
- Spectrum loading.
- Bolt clamping force.

The work conducted by Reemsnyder was very comprehensive and it is often cited by other researchers in the railway bridge field, despite the fact that the investigation was into such an "odd" construction as an iron-ore bridge. There are many similarities with a railway bridge though – and the fatigue loading is as easy, or perhaps easier, to define. As Wyly and Scott quite correctly pointed out in an article from 1956 [19]:

"These structures constitute, in effect, gigantic fatigue testing machines. The full-size specimens are tested repeatedly under actual conditions. The load, closely known in magnitude, is nearly constant hour after hour, year after year. The number of cycles of loading is also known to a reasonable degree of accuracy."

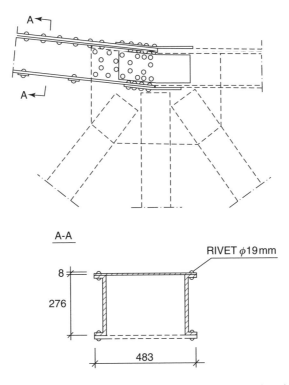

Fig. 2.1 A detail of the truss joint and a cross-section of the truss member that was fatigue tested by Reemsnyder in 1975 [18].

The riveted truss members from the iron-ore bridge are very similar in shape to railway truss bridge members, and as Reemsnyder himself pointed out, they can give an insight into riveted structures in general given that:

- There are a large number of old riveted structures still in service.
- These structures are also carrying a larger load than they were originally intended for.
- At the time these structures were designed, the effect of fatigue due to cyclic loading was often neglected, and if such an effect was taken into account, it was based on limited understanding and knowledge of the phenomenon.

The riveted truss joint that was selected for the fatigue testing and the adjoining truss member are shown in Fig. 2.1.

In ten of the truss members tested by Reemsnyder, he replaced the rivets in those critical locations where a fatigue crack could be observed. The fatigue testing of these members was then continued until complete failure of the section. A comparison of the fatigue life of those members that were reinforced with high-strength bolts and those that were not reinforced is shown in Fig. 2.2.

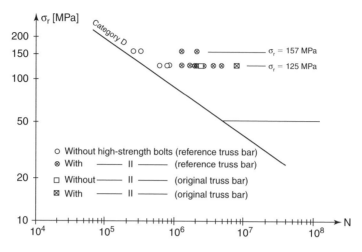

Fig. 2.2 The results from the fatigue tests conducted by Reemsnyder in 1975 [18] on riveted truss members taken from an iron-ore bridge. Two of the 18 members were subjected to spectrum loading and therefore could not be marked on this constants-amplitude diagram. In order to compare the results from Reemsnyder's investigation, the relevant fatigue design curve for riveted details given by AASHTO (American Association of State Highway and Transportation Officials), AREA (American Railroad Engineering Association) and ECCS (European Convention for Constructional Steel work) has been given in the diagram. Category D is the fatigue design curve given by AASHTO and AREA, and it is equivalent to the fatigue design curve for detail category 71 according to ECCS.

The results from the investigation conducted by Reemsnyder can be briefly summarized by two conclusions:

– When replacing the rivets in critical regions by high-strength bolts during the testing, there was in an increase in fatigue life of 2–6 times that of those members which had not been retrofitted (i.e. members where the rivets were not replaced).
– An increase in the clamping force of the high-strength bolts also resulted in a corresponding increase of the fatigue life of the truss members that were tested.

2.3.3 "Comportement à la fatigue de profiles laminés avec semelles de renfort rivetées" (Rabemanantsoa and Hirt 1984) [20]

In 1984 Rabemanantsoa and Hirt [20] tested four rolled railway bridge girders (HEB 1000) with riveted cover plates at the lower tension flange. The girders were intended for temporary use and had never been in service, and they were consequently in very good condition. They were tested by four-point bending (for more on four-point bending tests see Example 2 page 148) at relatively low stress ranges ($\sigma_r =$ 78–90 MPa) and therefore sustained up to $7 \cdot 10^6$ stress cycles. The fatigue cracks emanated exclusively from the rivet holes in the lower flange. The results are shown in Fig. 2.3.

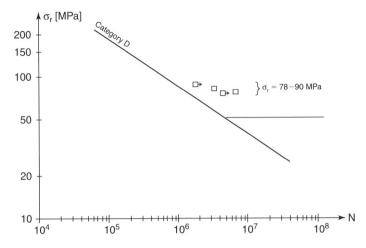

Fig. 2.3 The results from the fatigue test conducted by Rabemanantsoa and Hirt in 1984 [20] on four rolled girders with riveted cover plates at the lower tension flange. The fatigue design curve for riveted details (category D) is shown in the diagram for comparison.

Two of the girders were reinforced during the testing, and have been marked with an arrow symbol in the diagram. The marked load cycle value refers to the value that was reached just before the girders were reinforced. The two reinforced girders have an attached arrow in the diagram just to show that their fatigue life, without measures being undertaken, is longer. When the first fatigue crack was detected in the lower tension flange an extra cover plate was added to this flange, and if the crack then later propagated up into the web, the crack was arrested by drilling a stop hole at the tip of the crack. These measures were taken in order to simulate an actual reinforcing procedure in the field.

The length of the fatigue life of these two girders was more than three times that of the fatigue life of the unreinforced girders. The result of these fatigue tests shows that it is possible to considerably increase the service life of a fatigue-cracked girder bridge by reinforcing in a suitable manner.

2.3.4 "Fatigue strength of weathered and deteriorated riveted members" (Out, Fisher and Yen 1984) [21]

Out et al. performed a study in 1984 [21] on the fatigue resistance of four heavily corroded riveted built-up stringers taken from a bridge that was built in 1903 and demolished in 1982. Due to the fact that the stress ranges chosen were low and close to the fatigue limit of riveted details ($\sigma_r = 56$–75 MPa), the fatigue life was long (up to $40 \cdot 10^6$ cycles, see Fig. 2.4). The stringers were tested using a four-point bending test.

By drilling stop holes and given the fact that the stringers showed substantial redundancy (i.e. the ability of a riveted built-up I-section to carry load by "composite action" even after the complete failure of a section component), it was possible to continue the cyclic loading process even after a first failure. There were possibly also some additional reinforcing steps taken during the fatigue tests. This could, in that case, perhaps

14 Fatigue Life of Riveted Steel Bridges

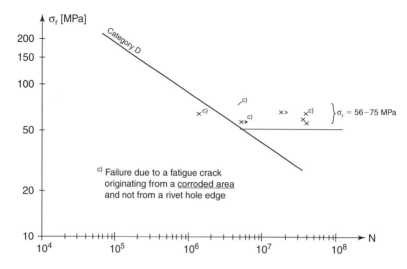

Fig. 2.4 The results from the fatigue test reported by Out, Fisher and Yen in 1984 [21] on four riveted built-up stringers. The fatigue design curve for riveted details (category D) is shown in the diagram for comparison.

explain why the authors show more results in Fig. 2.4 than the number of stringers tested.

Unless the corroded areas had not severely reduced the plate thickness before testing, there was not any crack initiation apart from cracks forming at the rivet holes. The test results of the investigation performed by Out, Fisher and Yen also showed that:

- A strong frictional bond between the section components was beneficial for the fatigue life.
- The built-up riveted I-section exhibited substantial redundancy after a crack had severed a component (e.g. a flange angle).
- The crack growth did not exhibit unstable propagation even though the temperature was reduced to −40°C at intervals during fatigue testing.

2.3.5 "Fatigue of riveted connections" (Baker and Kulak 1985) [22]

Baker and Kulak performed a series of tests in 1985 [22] to study the effect on fatigue life of replacing rivets with high-strength bolts. They simulated a riveted connection by merely using unfilled holes in the midspan of wide-flange beams subjected to four-point bending. The idea was to investigate the lower bound of riveted joint fatigue life (i.e. for a riveted connection with completely loose rivets). They pointed out that this procedure neglected both some beneficial and some detrimental effects that riveted connections are exposed to in service: clamping force, fretting and corrosion under the rivet head, and finally bearing stress – all were excluded. The results from these tests were then compared with those from a second series of similar tests, but now

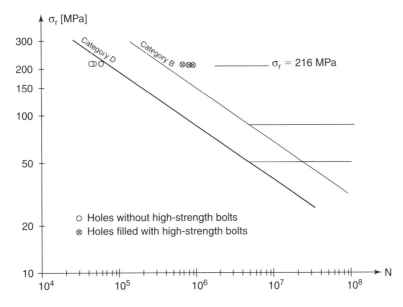

Fig. 2.5 The results from the fatigue tests conducted by Baker and Kulak in 1985 [22] on wide-flange beams with holes – unfilled and filled with high-strength bolts. The fatigue design curve for riveted details (category D) and the fatigue design curve for unnotched steel members (category B) are shown in the diagram for comparison.

with high-strength bolts in the holes. This resulted in up to 16 times longer fatigue life compared with the flanges with empty holes. These results are shown in Fig. 2.5.

In addition to these tests, Baker and Kulak also performed some full-scale tests on axially loaded riveted truss members taken from a highway bridge built in 1914. The fatigue strength of these actual riveted bridge connections proved to be higher than that obtained for the beams with empty holes and also higher than the fatigue strength specified in the design code (AASHTO: category D). These results are shown in Fig. 2.6.

By separating the lattice girder flanges from each other, and by subsequently shortening the length during the test, it was possible to obtain eleven fatigue life results from these four truss members.

2.3.6 "Fatigue and fracture evaluation for rating riveted bridges" (Fisher, Yen, Wang and Mann 1987) [24]

In 1987 Fisher et al. published the results from a fatigue test series on 14 riveted girders taken from three different bridges. The girders were all of the same type – riveted built-up I-sections. In order to reduce the bending stiffness, two of the girder sections were reduced in height and a new top flange was welded to the remaining web plate. This was done in order to decrease the fatigue load necessary to achieve the desired stress range. The result from the fatigue test is shown in Fig. 2.7. Only 13 fatigue results were obtained since one of the girders failed before the fatigue test (during a static load test).

16 Fatigue Life of Riveted Steel Bridges

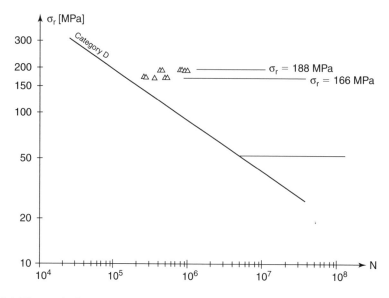

Fig. 2.6 The results from the fatigue tests conducted by Baker and Kulak in 1985 [22] on four riveted truss bridge members (diagonals and verticals – of the lattice girder type). The fatigue design curve for riveted details (category D) is shown in the diagram for comparison.

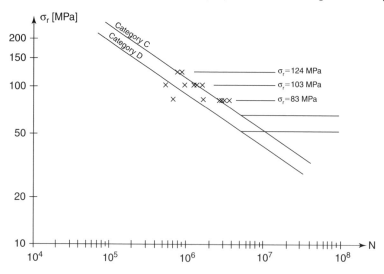

Fig. 2.7 The results from the fatigue test conducted by Fisher et al. in 1987 [24] on 13 riveted built-up girders. The fatigue design curves for riveted details, category D as well as category C, are shown in the diagram for comparison.

The main results from this investigation can be summarized as follows:

- The fatigue design curve for riveted details given in the codes (AASHTO: category D) could be used for approximate estimation of the fatigue life, although it is well on the safe side.

- The recorded fatigue life of the girders tested was closer to that of category C. (Note, the authors are probably referring here to the mean value.)
- For the three different stress ranges chosen for the test, three different mean stresses were used by randomly varying the minimum stress between 14, 55 and 96 MPa. Despite the great difference in mean stress, no correlation between mean stress and fatigue life could be determined.
- The fatigue test was performed from time to time under reduced temperature (periodic intervals at a temperature below $-40°C$ during the fatigue testing). The results showed that relatively large fatigue cracks can be sustained without any brittle fracture of a component.
- The riveted built-up I-girders exhibited a substantial inherent structural redundancy. A cracked section, where up to two section components had cracked, had the ability to redistribute load to other parts and components within the girder. This suggests that it should be easy to detect and take necessary measures in the time interval between the detection of large occurred cracks and a complete section failure.

2.3.7 "Fatigue strength of corroded steel plates from old railway bridge" (Abe 1989) [25]

In 1989 Abe tested nine riveted built-up stringers taken from an old railway bridge [25]. As well as undertaking some standardized fatigue tests on different details manufactured from steel samples taken from the bridge, he wanted to study the fatigue life of corroded parts and members that had been in service for a long period of time.

The stringers were in relatively good condition, except in the midspan region, where the tension flange was badly corroded at the point of a riveted connection for a horizontal wind bracing diagonal. Despite of this quite severe defect in the midspan region, one of Abe's principal findings was that the fatigue life of corroded riveted details is governed by the rivet hole rather than by the effect from corrosion, provided that the corrosion is not too severe. Abe's test results are shown in Fig. 2.8.

2.3.8 "Fatigue and fracture of riveted bridge members" (Brühwiler, Smith and Hirt 1990) [27]

In 1990 Brühwiler, Smith and Hirt [27] investigated the fatigue behaviour of six riveted built-up girders and three lattice girders in wrought iron removed from two road bridges built in 1894 and 1891 respectively. The girders were tested by four-point bending with constant-amplitude load cycling ($\sigma_r = 50-120$ MPa) with up to $20 \cdot 10^6$ stress cycles. The results are shown in Fig. 2.9.

The results were compared with those from the investigations performed by Rabemanantsoa and Hirt in 1984 [20]. It was found that wrought iron elements have fatigue strengths that are similar to that of mild steel riveted girders. Brühwiler et al. also found that corroded members do not show lower fatigue strength than non-corroded girders. They suggest that this may be due, in part, to the absence of corrosion at rivet holes. The test results also showed that a shear stress range of 100 MPa may be taken as an approximate constant-amplitude fatigue limit for rivets in shear.

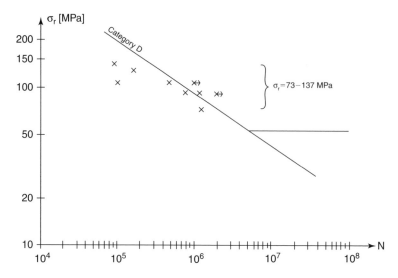

Fig. 2.8 The results from the fatigue test conducted by Abe in 1989 [25] on nine riveted built-up stringers. The fatigue design curve for riveted details (category D) is shown in the diagram for comparison.

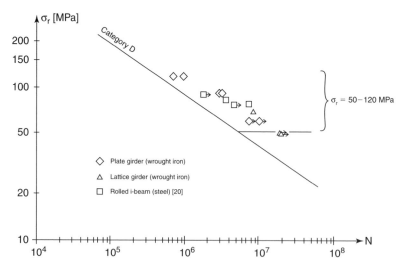

Fig. 2.9 The results from the fatigue test conducted by Brühwiler, Smith and Hirt in 1990 [27] on nine riveted wrought iron girders. The fatigue design curve for riveted details (category D) is shown in the diagram for comparison.

2.3.9 "Experimental and theoretical investigations of existing railway bridges" (Mang and Bucak 1990) [29]

Mang and Bucak published a paper in 1990 [29] that presented the results from a fatigue test series on a complete bridge section and some other bridge beams taken

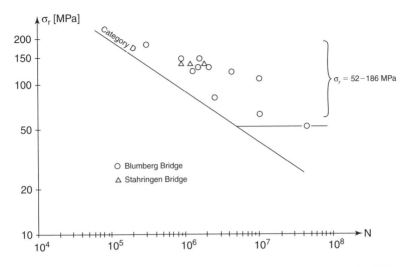

Fig. 2.10 The results from fatigue tests conducted by Mang and Bucak in 1990 [29] on riveted girders taken from two different railway bridges. The fatigue design curve for riveted details (category D) is shown in the diagram for comparison.

from two riveted railway bridges (the Blumberg Bridge and the Stahringen Bridge) built in the late 1800s. The results from the fatigue tests are shown in Fig. 2.10. For the Blumberg Bridge it was possible to obtain several fatigue life results by dismantling the bridge superstructure section into its constituent members after the first completed fatigue test, and then subsequently testing the members that were still intact.

The result from the tests showed fatigue lives in accordance to or above the fatigue strength curve for riveted details specified in the code. Mang and Bucak then conducted several other full-scale tests and this work was continuing at the time of this literature review (the early 1990s).

2.3.10 *A comparison and summary of the results from the full-scale fatigue tests*

This section compares and discusses the principal findings from the full-scale fatigue tests that have been summarized in sections 2.3.2–2.3.9. The investigation by Fairbairn in the 1860s [17] is excluded due to the fact that his only goal was to prove the deteriorating effect of fatigue on ultimate strength.

1. Most of the fatigue results achieved were in accordance with or above the fatigue strength curve for riveted details specified in the code, see Fig. 2.11.

   However, for some girders the fatigue life was shorter than that given by the fatigue design curve (category D). This was often due to severe corrosion damage in combination with bad detailing (e.g. horizontal wind bracing diagonals connected to the tension flange in the midspan region).

   A relevant question to ask here is whether it is correct to compare fatigue fracture values with a fatigue design curve? Such a curve should be used as a tool

20 Fatigue Life of Riveted Steel Bridges

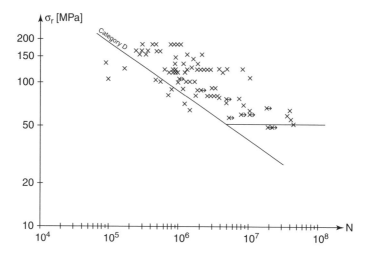

Fig. 2.11 A summary of all results from the different full-scale fatigue tests that have been presented in sections 2.3.2–2.3.9. The fatigue design curve for riveted details (category D) is shown in the diagram for comparison.

when designing *new* structures subjected to fatigue loading, but should it be used to make comparisons with actual fracture values? Is the category D design curve representative or not? Some comments that could be made in response to this last question are:

- By comparing the standard deviation for different stress ranges it is possible to estimate whether the category D fatigue design curve is representative for riveted details. It is easy to see that it is well on the safe side for the uncorroded test series considered here.
- If the category D design curve is representative, how can the fatigue results from different full-scale riveted members be above (rather than below) the curve given that the structures investigated have been in service for more than 50 years in many cases? One could reasonably assume that a considerable part of their fatigue life should have been "consumed" when subjected to more than 50 years loading.

This last remark leads to two conclusions: either the stress ranges during the years of loading have been *below* the fatigue limit, and so the curve is on the safe side as can be seen in Fig. 2.11, or the stress ranges have been *above* the fatigue limit, and in that case the fracture values given in the Fig. 2.11 should be compensated for fatigue damage accumulation – the category D design curve is then even more on the safe side.

2. There is a direct correlation between rivet clamping force and fatigue life. Reemsnyder [18] showed that by replacing the rivets in the critical regions with high-strength bolts the fatigue life is increased by 2–6 times in comparison to that of members where the rivets were left intact. Baker and Kulak [22] also showed

a substantial increase in fatigue life for girders with drilled holes in the lower tension flange containing high-strength bolts in high tension compared with girders with the fatigue life of unfilled holes.
3. Several investigations showed similar results concerning the impact of corrosion damage on the fatigue life of riveted details. If the corrosion damage was not too severe and if the rivet head had protected the rivet hole edge from any influence of corrosion, then the fatigue life was the same for riveted members with and without corrosion damage.
4. When studying the propagation of a fatigue crack, the existence of a substantial inherent structural redundancy was often pointed out. Through composite action a riveted built-up member has the ability to carry load even if one or two load-carrying member parts (e.g. L-flanges and/or flange plates) have cracked.
5. During the fatigue tests performed by Fisher et al. [24], the temperature was reduced at periodic intervals to $-40°C$ or below. No sign of any unstable crack growth (i.e. brittle fracture) was found, despite the fact that in general very low fracture toughness values have been noted when performing impact notch testing of the material.
6. No correlation between mean stress and fatigue life could be confirmed. Fisher et al. [24] carried out a factorial investigation during their fatigue tests. They let the mean stress vary between 56, 107 and 159 MPa. Despite the large variations in mean stress, no influence on the fatigue life could be ascertained.
7. No influence from fatigue damage accumulation could be determined. This was found by Abe [25], who performed a thorough analysis by comparing different parts with different "stress range histories".
8. The fatigue limit for rivets in shear (τ_{ru}) was found by Brühwiler et al. to be in the neighbourhood of 100 MPa. The investigations of the load-carrying capacity and the fatigue life of riveted railway bridges carried out by Åkesson, [1], [3] and [9], have shown that this tentative fatigue limit value is most probably correct. No calculated shear stress range in the bridges investigated by Åkesson has exceeded this value, and this could also explain why there are so few shear fatigue failures in rivets.

Almost all tests were performed at room temperature. The influence of (low) temperature on the fatigue life was only discussed in one of the cited works, namely Fisher et al. [24], cf. point 5 above.

Chapter 3

Riveted Connections

3.1 Historical review

Riveting was the dominant joining technique used in the construction of railway bridges in Sweden from the last decades of the 1800s until the late 1930s. Before that period, joints and connections were also fastened by the use of simple pin bolts.

When railway lines and railway bridges began to be built in Sweden in the late 1850s, the steel material was of the puddle iron type and it was mainly imported. From the beginning, however, the rivets were manufactured domestically. In the 1870s and 1880s, when the development and expansion of the railways in Sweden was at its peak, the material used for constructing bridges was Swedish steel. A very early example of a steel railway bridge was one built in 1866 near the city of Vänersborg. However, this bridge was not riveted but pin bolted.

The use of steel instead of iron combined with the use of rivets enabled the development of railway bridge structures with greater span lengths than before (such as riveted truss bridges). Until the turn of the century, the riveting of the different parts and members was always made by hand, but from 1900 onwards rivets were usually made by machine tools.

Even though riveting was simplified by making as much of the joint as possible in the workshops, the actual process was slow and required many labourers. However, the riveting technique was the dominant connecting method in railway bridge construction until the late 1930s. It was the introduction of welding techniques together with the use of concrete structures that gradually made rivets obsolete. Later in the twentieth century the use of high-strength bolts also removed the need for rivets. By utilizing bolts with greater strengths than rivets, it was not only possible to reduce the number of fasteners, but also to benefit from the positive effect on fatigue life resulting from the higher clamping force in high-strength bolted connections.

By the 1990s, the riveting technique had almost been forgotten and was no longer used when constructing new railway bridges in steel. The only time, however, when one had to utilize the old technique in railway bridge engineering was when there was a need to replace rivets that were loose or where the head had fallen off. These remedial actions were enough to keep the old technique alive, despite the fact that no riveted railway bridges were being built anymore.

24 Fatigue Life of Riveted Steel Bridges

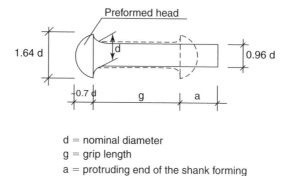

d = nominal diameter
g = grip length
a = protruding end of the shank forming
    the second head (~4/3 d+0.1g)

Fig. 3.1 A typical appearance of an undriven rivet.

3.2 Riveting technique

The unique and ingenious method of connecting steel plates and members together permanently by the use of the hot riveting process was the dominant connecting method in steel railway bridge construction in Sweden until the late 1930s. Although the method was labour-intensive and time-consuming the riveting process had no competitors, and we should bear in mind that at that time there was a large manual workforce and the labour cost was low.

The riveting process consists of three phases:

– First, the rivet head has to be heated to approximately 1000°C.
– While still glowing the rivet is then inserted into the hole, which should be at least one millimetre larger in diameter than the nominal diameter of the rivet in order to just be inserted easily enough.
– Finally, a second head (the first preformed head is on the other side) is formed by a riveting machine, which by rapid forging or by continuous squeezing moulds the protruding end, still plastic due to the remaining heat in the rivet, and normally the rivet shank then fills out the hole. This is shown in Fig. 3.1.

In the 1800s, the common method of making a rivet hole in a steel plate was by punching. This, together with hand-hammered rivets, often resulted in a poor quality riveted joint. In the beginning of the 1900s, after experiencing a number of rivet failures and early fatigue cracking emanating from the rivet hole edges, the railway authorities came up with new regulations and specifications concerning the riveting process and hole preparation methods, cf. refs. [114–117]:

– The rivet holes should be drilled. When the plate thickness is less or equal to 16 mm, a three millimetre sub-punching and then reaming to full diameter is allowed.

- The riveting procedure should be performed by using special riveting machines (pneumatic hammer and hydraulic or electric pressure riveter). This work should be performed in workshops wherever possible. Riveting work in the field should be kept to a minimum. Hand-hammered rivets should be used only in those places where riveting machines are impossible to use.
- In places where riveting becomes difficult to perform, the rivets should be replaced by bolts.
- The distance between the rivets (i.e. pitch and gauge) should be kept to a minimum in order to ensure that the matching predrilled plates will be tightly clamped to each other.
- Every rivet shall be freed from scale and glowing particles before being inserted into the hole.
- The rivets should completely fill the holes, and this should be controlled by removing some of the rivets after completing the riveting where necessary.
- The interfacing plate surfaces should be carefully cleaned by removing corrosion and other particles before being painted two times with a special kind of oil lead paint.
- Special care and attention should be taken when performing the riveting work, due to its importance for the structural performance.
- The riveting work should be checked by lightly tapping at the side of the rivet head with a small hammer. If the rivet is loose, it should be replaced immediately.

3.3 Load-carrying capacity

3.3.1 *Clamping force*

When the hot rivet has been inserted into the hole of the plates to be connected and the second head has been formed from the protruding shank, the rivet shortens in length due to cooling. However, most of the shrinkage of the free rivet is restricted by the connected plates which consequently are compressed. The residual tensile force in the rivet and the compressive force in the plates balance each other, i.e. the stresses are self-equilibrating.

The clamping force, applied by the rivet head is normally assumed to be evenly distributed through the plate thickness to the midplane with an angle α of 30–45° from the outermost part of the rivet head, see Fig. 3.2.

In 1961 Fernlund presented a theoretical solution [60], where he derived the lateral stress distribution at the midplane by assuming the plates acted as an "elastic soil" which deforms under the pressure from the rivet clamping force. In this model, the compressive stresses are the highest near the edge of the rivet hole and then gradually decrease in the radial direction.

The clamping force from the rivet generates a complex triaxial stress state in the connected plates in the vicinity of the rivet hole. Not only, as described above, are the plates being compressed laterally, but there are also radial and circumferential ("hoop") stresses. When the connected plates are compressed together by the rivet, the plate material in this region wants to expand in the radial direction. This expansion is restricted by "hoop rings", which are compressed near the rivet hole, and expanded in the outer part of the axially compressed region, see Fig. 3.3.

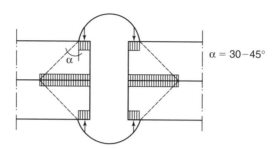

Fig. 3.2 The distribution of the compressive stresses through the plate thickness to the midplane – the normally accepted and simple model.

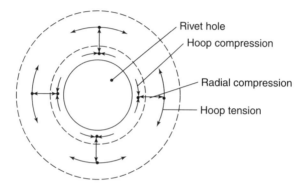

Fig. 3.3 Radial and circumferential "hoop" stresses near the rivet hole due to a clamping force from the rivet.

Wilson and Thomas performed a series of tests in 1938 [101] to measure the initial clamping force in rivets. They compared the effect of grip length and riveting method on the initial clamping force. The results from these tests show that the initial tension in the rivet is approximately 70% of the yield stress independent of the riveting method, see Table 3.1.

From the results in Table 3.1 one can see that the clamping force is increased as the grip length is increased. The average clamping ratio for the two grip lengths investigated is 0.61 and 0.77 respectively (75/125 mm). This fact is easily understood if one takes into consideration that the longer the rivet shank, the more it wants to contract when cooling after the riveting has been completed. But, as explained above, this contraction is restricted by the compressed plates and therefore, as the thickness of the connected plates is increased, the relative shortening of the rivet is decreased (i.e. the relative lengthening which is forced upon the rivet is increased). This is due to the fact that, as the thickness of the plates is increased, the ratio between the axial stiffness of the plates and the axial stiffness of the rivet also increases. In other words, considering the load distribution shown in Fig. 3.2, this conclusion can be summarized as follows: when the grip length is increased the compressed contact area between the plates also increases. It may be said that when increasing total plate thickness the reducing effect

Table 3.1 The relative stress σ_i/σ_y, i.e. the ratio between the initial tensile stress and the yield stress in hot driven carbon steel rivets (data compiled and collected from the investigation by Wilson and Thomas in 1938 [101]).

Specimen No.	Rivet Diameter ϕ (mm)	Grip Length L (mm)	Riveting Method	Yield stress σ_y (MPa)	Initial Tension σ_i (MPa)	'Clamping ratio' σ_i/σ_y
R1-1	25	75	Pneumatic hammer	316	214	0.68
R1-2	25	75	Pneumatic hammer	337	163	0.48
R1-8	25	125	Pneumatic hammer	314	238	0.76
R1-9	25	125	Pneumatic hammer	298	235	0.79
R3-1	25	75	Hydraulic machine	346	227	0.66
R3-7	25	125	Hydraulic machine	323	245	0.76
R3-9	25	125	Hydraulic machine	319	248	0.78

Average: 0.70

on the axial stiffness is counterbalanced by an increased area. The rivet does not have this "compensating effect" when increasing in length. Studies have been made of this difference in contraction for different plate thicknesses. It has been observed that when the grip length is increased there is a gap between the plate hole edge and the rivet shank, which increases as the shank length increases. As the grip length is increased, and the relative elongation of the rivet is increased, the lateral "Poisson type" deformation of the rivet shank is also increased. Whether or not these studies have taken into consideration the compensating length of the protruding end of the shank (0.1 x grip length – see Fig. 3.1) is not known.

By using the results from Table 3.1 (taking specimen R1-1 as an example) we can calculate the magnitude of both the clamping force F and the resulting elongation ΔL of the rivet (equations 3.1, 3.2):

$$F_{clamp} = \sigma_i \cdot A_{rivet} = 214 \cdot 10^3 \cdot 491 \cdot 10^{-6} = 105 \, kN \tag{3.1}$$

$$\Delta L_{rivet} = \frac{F \cdot L}{E \cdot A} = \frac{105 \cdot 75 \cdot 10^{-3}}{2.1 \cdot 10^8 \cdot 491 \cdot 10^{-6}} = 0.076 \, mm \tag{3.2}$$

Although the clamping force is high, the corresponding axial elongation of the rivet is extremely small. The deformation in the rivet can also be derived using the simple expression shown in Fig. 3.4.

Consider a simple stress distribution due to a clamping force defined by an angle α of 45°, see Fig. 3.5. By assuming a plate ring with an average compressed diameter D, one can make a simplified estimate of the axial stiffness of the plates.

Taking once again specimen R1-1 from Table 3.1 as an example, the results are as follows (for $d = 25$ mm and $L = 75$ mm) (equation 3.3):

$$D = 77.5 \, mm \Rightarrow A_{plate} = 4226 \, mm^2 \, (A_{rivet} = 491 \, mm^2) \tag{3.3}$$

$$\Delta L_{rivet} = \Delta L_{temp} - \Delta L_{plate}$$

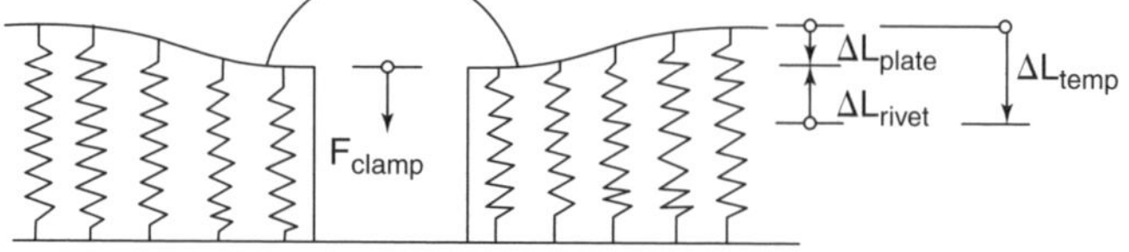

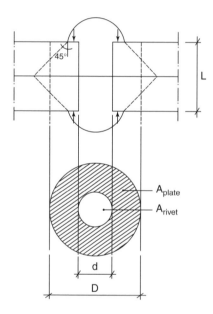

ΔL_{rivet} = Residual elongation (due to restricted contraction) of the rivet
ΔL_{temp} = Total thermal contraction (unrestricted condition)
ΔL_{plate} = The compression of the connected plates

Fig. 3.4 The deformation in a plate and rivet due to a clamping force.

Fig. 3.5 The average compressed width D of the plates for an assumed distribution angle $\alpha = 45°$.

The tensile force in the rivet should be balanced by the compressive force in the plates (equation 3.4):

$$F_{rivet} = \Delta L_{rivet} \cdot \left(\frac{E \cdot A}{L}\right)_{rivet}$$

$$F_{plate} = \Delta L_{plate} \cdot \left(\frac{E \cdot A}{L}\right)_{plate}$$

$$\Rightarrow F_{rivet} = F_{plate} \qquad (3.4)$$

Both the rivet and the equivalent cylindrical plate ring have the same stress-free length and modulus of elasticity, therefore expression (3.4) becomes:

$$\Delta L_{rivet} \cdot A_{rivet} = \Delta L_{plate} \cdot A_{plate} \tag{3.5}$$

The deformation in the plates for specimen R1-1 is:

$$\Delta L_{plate} = \Delta L_{rivet} \cdot \frac{A_{rivet}}{A_{plate}} = 0.076 \cdot \frac{491}{4226} = 0.009\,mm \tag{3.6}$$

The deformation in the plates could also be expressed as a function of the unrestricted thermal contraction and the different stiffnesses (i.e. in this case the different areas):

$$\Delta L_{plate} = \frac{\Delta L_{temp} \cdot A_{rivet}}{(A_{rivet} + A_{plate})} \tag{3.7}$$

This expression (3.7) gives a good explanation of the behaviour observed when analyzing the extreme limits of the plate stiffness:

$$A_{plate} = 0 \Rightarrow \Delta L_{plate} = \Delta L_{temp} \tag{3.8}$$
$$A_{plate} = \infty \Rightarrow \Delta L_{plate} = 0 \tag{3.9}$$

- The first extreme limit (3.8) is representing a totally unrestricted condition, i.e. where the rivet is fully allowed to contract without any induced internal stresses forced upon it.
- The second extreme limit (3.9) could be expressed as follows: as the relative stiffness of the plates is increased (i.e. as the grip length is increased), more residual stress (elongation) is forced upon the rivet when cooling.

By using the expression given in Fig. 3.4, one can calculate the value of the unrestricted thermal contraction:

$$\Delta L_{temp} = \Delta L_{rivet} + \Delta L_{plate} = 0.076 + 0.009 = 0.085\,mm \tag{3.10}$$

This thermal contraction is equivalent to a temperature drop of about 100°C:

$$\Delta L_{temp} = \alpha \cdot t \cdot L = 12 \cdot 10^{-6} \cdot 100 \cdot 75 = 0.09\,mm \tag{3.11}$$

Taking into consideration the fact that the deformations in general are extremely small and the driving temperature during riveting is much higher than this value, and putting aside any doubts about this simplified model, one can draw the following conclusions:

- The surface finish and surface treatment is of great importance when producing clamping forces of a controllable size.
- The resulting clamping force is directly dependent on the finishing driving temperature. The lower the driving temperature, the lower the resulting clamping forces.

30 Fatigue Life of Riveted Steel Bridges

Table 3.2 Requirements for the rivet and plate material as given by the design specification for the Forsmo Bridge (built in 1912).

	Tensile strength (MPa)	Elongation (%)	Chemical content (%)
Rivet material[1]	315–370	≥ 28[2]	–
Plate material	385–440	≥ 22[2]	$P \leq 0.07$

1) That is, undriven rivets.
2) Measured on a total length of 20 cm.

Wilson and Thomas conducted their series of tests under controlled laboratory conditions and therefore the scatter in the results in Table 3.1 showing the clamping force is low. The experience in normal conditions, and especially from field practice, is that the resulting clamping force can vary substantially and that consequently one can not set a reliable value to be used in design. Furthermore, one can assume a certain relaxation of the rivet clamping force due to creep and to fretting of the interfacing plate surfaces, cf. section 3.4. These factors have an uncertain magnitude but one thing is sure, their influence increases over time (i.e. as the bridge gets older!).

3.3.2 Mechanical properties

For older riveted railway bridges in Sweden, it is sometimes possible to find the original design specifications. These give the physical specifications of the steel material. In a thorough manner, they describe the allowable limits concerning the mechanical properties as well as the chemical content. As an example, Table 3.2 shows the specifications issued in 1912 when the Forsmo Bridge was built. At that time, the Forsmo Bridge was the largest riveted railway bridge in Sweden (maximum span length 104 m, total length 263 m). For further data about the bridge see ref. [1].

Table 3.2 shows that there are lower tensile strength requirements and higher deformability demands for the rivet material than for the plate material. The demands for a higher level of deformability in the rivet material than in the plate material most likely were made because of these three circumstances:

1. During riveting, the rivet is subjected to a work hardening process by the mechanical action of the hammer. In addition, there is probably an increase in the grain size of the rivet material due to its heating and subsequent slow cooling. These two factors, work hardening and the increase in grain size, both contribute to a decrease in the ductility of the driven rivet compared with that of the undriven rivet material.
2. Due to the residual elongation of the rivet after riveting, there is reduced strain ductility for rivets subjected to tensile forces.
3. Rivets in shear (not taking a possible clamping force into consideration) are normally accompanied by large shear deformations, in contrast to the normally very small axial elongations caused by a pure tensile force. This is due to the bending of the shank that takes place under shear forces.

Table 3.3 Compiled and collected data from the investigation in 1938 by Wilson and Thomas [101] concerning changes in mechanical properties for carbon steel rivets.

Specimen No.	Grip length (mm)	Yield point σ_y (MPa)	Tensile strength σ_B (MPa)	Elongation[1] A (%)	Riveting method	
R1-1	75	316	477	28.0	Pneumatic hammer	Average for undriven rivet rod material:
R2-2	75	337	481	22.0	Pneumatic hammer	
R1-8	125	314	459	20.8	Pneumatic hammer	
R1-9	125	298	457	21.2	Pneumatic hammer	$\sigma_y = 265$ MPa
R3-1	75	346	481	32.0	Hydraulic machine	$\sigma_B = 403$ MPa
R3-7	125	323	468	24.6	Hydraulic machine	A = 36.1%[2]
R3-9	125	319	465	21.8	Hydraulic machine	
	Average:	322 MPa	470 MPa			

1) Measured on a length equal to the grip length.
2) Measured on a length of 50 mm.

The mechanical treatment of a rivet during riveting resembles a wrought iron (steel) process – in almost the same manner, the red-hot rivet is mechanically worked until the second head is formed from the protruding end of the shank. During this process the rivet's mechanical properties change – there is an increase in strength accompanied by a decrease in ductility. During the series of tests that Wilson and Thomas performed in 1938 [101], in which they studied the initial clamping force (see section 3.3.1), the changes in the mechanical properties of rivets were also investigated. Some of these results are presented in Table 3.3.

The results in Table 3.3 can be summarized as follows:

– The average yield point increased by 22%.
– The average tensile strength increased by 17%.
– There is a clear evidence of a reduction in ductility even though the elongation was measured on different lengths.
– No significant difference in the results between riveting methods can be noted.
– There is a noticeable difference in results for the two different grip lengths. The increase in yield point and tensile strength is greater for the shortest grip length:

Grip length	Yield point increase	Tensile strength increase
75 mm	+26%	+19%
125 mm	+18%	+15%
Average:	+22%	+17%

This last comment on the results in Table 3.3 is worth further explanation. Due to lateral plastic expansion of the hot rivet shank during riveting, one can assume that the plate hole edge is generating a lateral compression by restricting this "Poisson type" expansion. The more this lateral expansion is restricted, the more one could assume an increase in the residual strength of the rivet. For short grip lengths there is only a small "constraining action" from the steel plates, which restricts the subsequent lateral thermal movement of the rivet, and (as explained in section 3.3.1) when the grip length is decreased, there is a tendency for the rivet shank to contract laterally.

Another aspect that affects the tendency for the hole edges to restrict this lateral expansion is the hole clearance. In Sweden the rivet holes are generally oversized by one millimetre, while in the United States the same gap is 2.5 mm (0.1 inch). Smaller hole clearances require more planning and matching when connecting a multi-riveted joint. But the advantage is that one obtains a more rigid joint. The clamping force could also be assumed to slightly increase as the hole clearance is decreased. (When the effective contact pressure area under the rivet head is increased, there is an increased restricting "area" to withstand the thermal contraction from the rivet.)

To sum up this discussion about hole clearance and its influence on different parameters, as the hole clearance is decreased there is evidence to suggest that the joint's strength, as a result of riveting, is increased.

And what about changes in the mechanical properties of the surrounding plate material in the vicinity of the rivet hole? Although no studies have been found concerning this question, there is reason to believe that the plate material is largely unaffected during the riveting process. The mechanical treatment is strictly confined to the rivet shank and the heating of the joined parts has negligible effect.

3.3.3 Shear strength

A riveted joint subjected to shear is designed on the basis of the bearing strength or shear capacity of the rivet. In reality, taking the clamping force into account, shear loads of moderate magnitude are transferred by friction, just as for friction-grip joints. These are the primary reasons for not utilizing this "frictional behaviour" in static design:

- In contrast to the situation for friction-grip bolts, the actual riveting process does not provide a controlled clamping force in the joint.
- The residual clamping force in a riveted joint varies in magnitude from one rivet to another. This is due to a multitude of factors that influence the clamping force, including the driving temperature, grip length, riveting technique, the evenness and plane tolerance of the plates to be connected, and the performance of the riveting personnel.
- When comparing the clamping force in a riveted joint and in a high-strength friction-grip bolted joint, one finds a large difference in magnitude for several reasons. A friction-grip bolt is pretensioned to about 70% of the *tensile strength*, while in normal rivets the average value is about 70% of the *yield point* (cf. Table 3.1). The tensile strength for a typical friction-grip bolt is about 800 MPa or more, and the yield point for a driven rivet of ordinary mild carbon steel was found to be about 320 MPa (cf. Table 3.3). Furthermore, the evenness of the plate surface in a riveted joint is not as strictly controlled as in friction-grip bolted joint. If the connected plates are uneven, the residual clamping force is also a function of the (local) flexibility of the plates (due to the small gaps) and not only of their "axial" stiffness.
- The shear capacity of a friction-grip bolted joint depends both on the clamping force and on the friction coefficient (and, for multiple plate connections, also on the

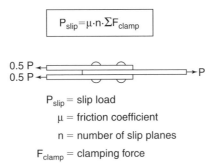

Fig. 3.6 The slip load for a riveted double-lap joint is (for moderate shear loads) a function of the friction coefficient, the number of slip planes and the total clamping force.

number of plates). In order to ensure that only these parameters are governing the slip shear capacity, the hole diameter is deliberately oversized a great deal (≥ 2 mm), and special care and attention is given to the surface finish and treatment of the plates. In riveted joints there is a markedly smaller hole clearance (due both to a smaller initial oversizing of the hole and to the expansion of the rivet shank as a result of riveting) and less strict demands placed on the plate finish, which results in a transfer of shear loads by a combination of bearing and friction between the connected plates. A transfer of shear loads through bearing in a friction-grip bolted joint is not considered in design, and it is also avoided by the large hole clearance combined with the high clamping force. Slip deformations of the same magnitude as the hole clearance (≈ 1 mm) are not tolerated in a joint that is designed to be stiff and rigid. Friction-grip bolted joints are sometimes used in combination with welded joints, and this is possible due to the large rigidity of friction-grip bolted joints. A riveted joint is more flexible than a high-strength friction-grip bolted joint.

All these reasons lead to the conclusion that the design recommendations, where riveted joints are treated as pure shear/bearing joints, are justified. However, the findings concerning the shear capacity of a friction-grip bolted joint (and its ability to transfer load through friction) are often valid for riveted joints subjected to moderate shear loads. Taking once again specimen R1-1 as an example (see Table 3.1 and also the calculation of the clamping force), it is possible to estimate the slip load for a simple double-lap joint, see Fig. 3.6.

The slip load for the double-lap joint shown in Fig. 3.6, with two rivets in a row and with $F_{clamp} = 105$ kN, can then be calculated:

$$\begin{aligned} \mu &\approx 0.33 \\ n &= 2 \\ F_{clamp} &= 105 \, kN \end{aligned} \Rightarrow P_{slip} = 0.33 \cdot 2 \cdot (2 \cdot 105) = 139 \, kN \quad (3.12)$$

This simple calculation clearly indicates that a riveted connection is capable of transferring a substantial amount of shear load by friction between the planes before

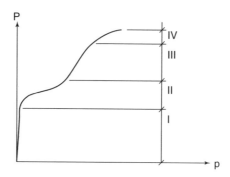

Fig. 3.7 Joint load *P* versus slip deformation *p* in a riveted joint (ideal conditions) [40], cf. Fig. 3.6.

slipping. This is an important observation, not only for moderate static shear loads but also for repeated loading. The expected fatigue life of a riveted joint is highly dependent on whether the shear force is transferred by friction or by bearing (see section 3.4).

Even though the calculation above shows that a substantial amount of shear load can be transferred by friction in a riveted joint, the ultimate strength of the joint is a function of the bearing capacity (for rivet or plate) or the shear strength of the rivet. Fig. 3.7 shows the characteristic load-deformation behaviour of a typical riveted joint.

Stage I   The slip is prevented by friction, and consequently the joint is very rigid.
Stage II Here the load is greater than the static friction, and the connected plates slip. Eventually the rivets come into bearing.
Stage III In this stage, the rivet shank and the plate hole edge deform elastically.
Stage IV In this final stage, the ultimate strength is a function of either the bearing capacity or the rivet shear strength. A substantial amount of yielding results from this excessive loading [40].

The number of slip planes is a governing parameter irrespective of whether the ultimate shear strength of a riveted lap joint is governed by the bearing capacity or by the shear strength of the rivet:

Number of slip planes   Ultimate governing capacity (for normal dimensions)
1   Shear strength of the rivet
≥ 2 Bearing capacity of either plate or rivet

The ultimate tensile strength of a lap joint is practically independent of factors that influence the initial stiffness of the joint or other deviations which influence the initial load-carrying behaviour (see Fig. 3.8). Factors such as the variation in clamping force, bending deformations in rivets due to long grip lengths, and misalignment, all contribute to the behaviour under a small load *P*, but do not affect the ultimate strength of the joint.

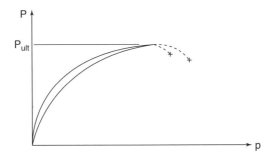

Fig. 3.8 The ultimate shear strength of a lap joint is unaffected by different factors that influence the initial stiffness of the joint.

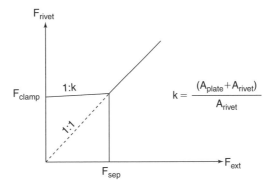

Fig. 3.9 Tension in the rivet as a function of an externally applied tensile force.

3.3.4 Tensile strength

Because of the prestressing effect, which is a consequence of the residual clamping force in the connection, the result of an external tensile load will primarily be a loss of lateral compression between the plates – very little load is initially carried by the rivet. This is due to the great difference in axial stiffness between the compressed part of the plates and the rivet. As previously described in section 3.3.1, there is a self-balanced system of internal stresses that enables the connected plates to carry tensile forces in the lateral direction, and, because of the great stiffness of the plates in comparison to the rivet, the plates will carry the main part of an external tensile load until a separation of the plates takes place. This behaviour is graphically explained in Fig. 3.9.

As can be seen in the schematic graph in Fig 3.9, the tension in the rivet only increases slightly for small external loads. If we once again use the results from the calculations in section 3.3.1 (for specimen R1-1), we find that:

$$A_{plate} = 4226 \, mm^2$$
$$A_{rivet} = 491 \, mm^2 \Rightarrow k \approx 10 \qquad (3.13)$$

36 Fatigue Life of Riveted Steel Bridges

In this example, an external load is distributed between the plates (as a reduction in compressive stress) and the rivet (as an increase in tensile stress):

$$\Delta F_{rivet} = \frac{1}{k} \cdot F_{ext} = 0.1 \cdot F_{ext} \tag{3.14}$$

$$\Delta F_{plate} = \left(1 - \frac{1}{k}\right) \cdot F_{ext} = 0.9 \cdot F_{ext} \tag{3.15}$$

Thus only 10% of the increase in the external load is distributed to the rivet; the rest (i.e. 90%) is taken by the compressed plates. If on the other hand we consider an example where the plate thickness is smaller than that of specimen R1-1, there would be an increase in the inclination of the first part in the curve in Fig. 3.9. A decrease in plate thickness (i.e. a decrease in plate stiffness) would result in a decrease of the value k — so a greater part of an external load would be taken by the rivet.

For a given value of the external load ($F_{ext} = F_{sep}$), the connected plates begin to separate and the load-carrying capacity of the plates is then reduced to zero – the rivet is taking 100% of the external load. This point of separation can easily be derived:

$$F_{rivet} = F_{clamp} + \frac{1}{k} \cdot F_{sep}$$

$$\Rightarrow F_{sep} = \frac{F_{clamp}}{\left(1 - \frac{1}{k}\right)} \tag{3.16}$$

$$F_{rivet} = F_{sep}$$

For the value of k equals to 10, the "separation load" F_{sep} would then be approximately 11% above the clamping force:

$$F_{sep} = 1.11 \cdot F_{clamp} \tag{3.17}$$

As this expression clearly shows, a decrease in the clamping force will result in lower separation load, and consequently a substantial loss of joint stiffness at an earlier stage due to the loss of load-carrying capacity in the plates. The ultimate tensile capacity of the joint is, however, only a function of the tensile strength of the rivet and is therefore unaffected by the initial value of the clamping force.

Knowing that the ultimate tensile strength of a joint is unrelated to the clamping force one may ask what are the advantages of any prestressing behaviour? There are basically two beneficial effects at moderate loading levels ($F_{ext} < F_{clamp}$) compared to a joint in an unclamped condition:

- An externally applied tensile load will mainly be carried by the connected (and compressed) plates. Very little load is carried by the rivet. Denoting the stiffness of plate and rivet by S, we can write:

$$F_{ext} = S_{plate} \cdot p + S_{rivet} \cdot p \tag{3.18}$$

Riveted Connections

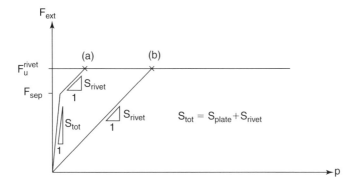

Fig. 3.10 The load-displacement characteristics for (a) a riveted joint with a normal level of clamping force and (b) a riveted joint in a totally unclamped condition.

$$k = 10 \Rightarrow F_{plate} = 0.9 \cdot F_{ext} \tag{3.19}$$

– The load-displacement characteristics are governed by the connected plates, and due to superior stiffness relative to the rivet, the resulting deformation is very small, cf. Fig. 3.10.

As Fig. 3.9 shows, for moderate loading levels ($F_{ext} \leq F_{clamp}$) the major part of the external load is taken by the plates. However, as we have noted, there is a tendency for the rivet to carry a greater portion of the load for thinner plates. In addition to an increase in the rivet force as the plate thickness is decreased, there could also be some secondary forces due to prying, which furthermore amplifies the tension in the rivet. The negative effect of prying is accentuated for riveted connections consisting of L- or T-shaped profiles, where the connected legs are flexible and therefore introduce additional tensile forces in the rivet due to a compressive reaction force in the outer part of the leg. The ultimate tensile strength of a riveted joint can be significantly reduced by the effect of prying. The principal effect of prying on a riveted joint is shown in Fig. 3.11.

Fisher and Struik [11] report that a difference in clamping force does not affect the prying forces at the ultimate loading stages. This again suggests that the tensile strength of the joint is ultimately unaffected by the clamping.

In a full-scale structure, such as a riveted truss railway bridge, there are numerous places where riveted L-shaped joints experience tension and possibly also a prying action:

– The web connection between the stringer and the floor beam in the track superstructure.
– The rigid moment connection between the floor beam and the truss (i.e. the connection to the vertical truss bars).
– The compression flange of the stringers (subjected to rotational bending due to eccentric loading from the sleepers).

38 Fatigue Life of Riveted Steel Bridges

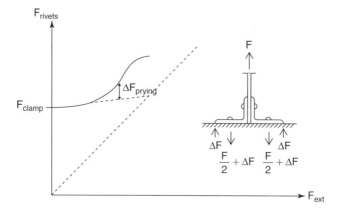

Fig. 3.11 The tension force in the rivet as a function of an external tensile force applied on an L-profile web connection where *prying* is a consideration (principal behaviour, after some results presented by Fisher and Struik [11]).

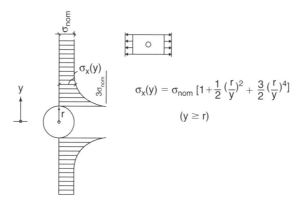

$$\sigma_x(y) = \sigma_{nom}\left[1+\frac{1}{2}\left(\frac{r}{y}\right)^2 + \frac{3}{2}\left(\frac{r}{y}\right)^4\right]$$

$(y \geq r)$

Fig. 3.12 The stress distribution in an infinite plate with an open hole, loaded axially with an evenly distributed tensile force.

– The vertical web stiffeners of the stringers where the horizontal and vertical cross-framing bars are attached.

3.4 Fatigue life of riveted connections

The most severe defect (i.e. stress raiser) in a riveted connection, with respect to fatigue, is the rivet hole itself. For an axially loaded infinite plate with a centrally located hole, the nominal stress at the net section is raised by a factor of three at the edges of the hole, see Fig. 3.12.

The stress concentration factor K is defined as the ratio between the maximum stress at the defect (in this case the rivet hole) and the nominal stress at the gross section. For the example shown in Fig 3.12, the stress concentration factor K is equal to 3. Even though there is substantial increase in the stress near the rivet hole edge, the stress is nearly of the same magnitude as the nominal stress level at a distance of approximately the rivet hole diameter from the rivet hole edge.

The increase in the nominal stress near the rivet hole edge together with small "micro-defects" originating from the fabrication process, such as small micro-cracks, flaws and notches, could be enough to produce small irreversible (i.e. plastic) strain deformations, even though the nominal stress level is well below the elastic limit of the material. Irreversible dislocation movements within the material could then produce a propagating (fatigue) crack if the stress reversal (i.e. stress range) is of a certain magnitude and is repeated often enough. As plastic deformations occur, the strain ductility is gradually decreased, and the material will become harder and more brittle, and consequently more susceptible to crack initiation. The material behaves in the same manner both when the deformations are global and when they are concentrated at a micro-level area. In addition to the embrittlement due to the plastic strain deformations, there is also some initial coldwork damage from the drilling (or, even worse, punching) of the rivet hole, which can contribute to a fatigue crack initiating even earlier.

If the stress range is high enough and is repeated a sufficient amount of times, the fatigue crack propagates until the net section tension capacity is ultimately reached. At this moment, the loaded plate element will experience a final and total fatigue failure. For the example in Fig 3.12, where a fatigue crack is initiated at the rivet hole edge, the crack propagates in a transverse direction to the longitudinal stress and is continuously reducing the effective net section area.

A single plate with an open hole could be seen as the case giving the lower bound of the fatigue life of a riveted connection. If the connection is non-bearing (i.e. there is no transfer of shearing forces) and the rivet clamping force is neglected, the stress concentration factor K is also about 3 (as is the case in Fig. 3.12), and consequently the fatigue resistance (or susceptibility) is the same. If on the other hand we consider a riveted connection (once again neglecting the effect of the clamping force) where there is a transfer of a shear force by bearing stresses (such as in a double lap joint – cf. Fig. 3.6), the stress concentration factor is increased due to the concentrated load transfer at the hole edge. For a double lap joint with only one rivet, the stress concentration factor K is usually equal to or above 5 depending on the bearing ratio. This ratio is defined as the nominal bearing stress divided by the nominal stress in the plate gross section. The fatigue life of a riveted connection has been found to be highly dependent on the bearing ratio. For a single rivet lap joint it is easily shown that the bearing ratio is only a function of the plate width b and the rivet hole diameter d:

$$\sigma_{bearing} = \frac{F}{d \cdot t}$$
$$\sigma_{nom} = \frac{F}{b \cdot t}$$
$$\Rightarrow bearing\ ratio = \frac{\sigma_{bearing}}{\sigma_{nom}} = \frac{b}{d} \qquad (3.20)$$

For this case, the bearing ratio is increased when the rivet hole diameter is decreased. The reason why the fatigue life is shorter when the bearing ratio is increased (i.e. when the rivet diameter is decreased) is that there is a more concentrated load transfer (on a smaller contact pressure area) and also an increase in local distortion around the hole.

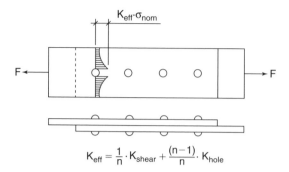

Fig. 3.13 The effective stress concentration factor in the plate net section at the first rivet location.

The bearing pressure also induces some "Poisson type" deformations in the plane of the plate perpendicular to the load direction (due to the ovalization of the hole), which are somewhat restricted and therefore introduce transverse tensile stresses near the rivet hole edge at the plate net section. These distortional stresses in combination with the increase in longitudinal peak stress explain why the fatigue life of a riveted shear connection is lower than that for an identical connection where no shear forces are transferred.

For a multi-riveted shear connection, however, the stress connection factor is decreased. This is because of the flow of stresses, which distributes the applied force to the different rivets. Consider the simplified model presented by Yin et al. [113] exemplified in Fig. 3.13 for the case of a four-row riveted single lap joint.

For a four-row rivet connection ($n=4$) the effective stress concentration factor becomes:

$$K_{shear} = 5$$
$$\Rightarrow K_{eff} = \frac{1}{4} \cdot 5 + \frac{(4-1)}{4} \cdot 3 = 3.5 \quad (3.21)$$
$$K_{hole} = 3$$

The calculation of the effective stress concentration factor in this example can be explained as follows: one fourth of the tensile force is transferred by shear by the first rivet. The rest of the force is transferred to the other rivets, and it therefore passes the first rivet location without producing any additional stresses other than those due to the rivet hole.

Note that as the number of rivets increases, the effective stress concentration factor approaches the value 3 (equal to K_{hole}):

$$n \rightarrow \infty \Rightarrow K_{eff} \rightarrow K_{hole} = 3 \quad (3.22)$$

This behaviour indicates that increasing the number of rivets in a row increases the fatigue resistance (lifespan). This statement, however, does not take into consideration the fact that when the total plate length is increased there is a tendency for the outer rivets to carry a greater proportion of the shear load due to plate elongation. The stress

Riveted Connections

concentration factor at the outer rivet will therefore not be reduced to the same extent as is suggested by the simplified model of Fig. 3.13.

The clamping force from the rivets, which exerts a lateral pressure to the plates at the rivet locations, has so far been neglected in this discussion concerning fatigue life. The clamping pressure is concentrated in the region under the rivet head and it has a positive effect on the fatigue resistance:

- First, it prevents the joined plates slipping under shear loading. For moderate load levels, the shear force is instead transferred mainly by friction rather than bearing (although the real behaviour is, in normal cases, probably a combination of both). The stress concentration near the edges of the rivet hole is therefore markedly reduced and, consequently, the stress in the net section is more evenly distributed (which reduces localized deformations at the rivet hole).
- Second, the localized clamping pressure around the rivet hole generates a favourable triaxial compressive stress state (compression laterally, circumferentially and radially – see Fig. 3.3). These compressive stresses will delay the initiation of a tension crack.

The positive effect on fatigue life from the clamping force can be illustrated by the following simplification, which is based on the extreme case of a riveted joint without transfer of shear:

Limit	Clamping force	Equivalent detail model	
Lower bound	0		A plate with a circular hole
Upper bound	$1.0 \cdot \sigma_y \cdot A_{rivet}$		An intact plate

The detrimental effect on the fatigue life due to a circular hole in a plate is thus decreased when there is a clamping force from a rivet.

If we once again use the results for specimen R1-1 (see Table 3.1 and the calculations previously made in this chapter) we can calculate the magnitude of the clamping pressure directly under the rivet head (see also Fig. 3.1):

$$F_{clamp} = 105 \, kN$$

$$\Delta A = A_{head} - A_{rivet} = \frac{\pi \cdot (1.64 \cdot 25)^2}{4} - 491 = 829 \, mm^2 \quad (3.23)$$

$$\Rightarrow \sigma_{clamp}^{plate} = \frac{F_{clamp}}{\Delta A} = \frac{105 \cdot 10^{-3}}{829 \cdot 10^{-6}} = 127 \, MPa \quad (3.24)$$

Carter, Lenzen and Wyly [59] found that the circumferential (hoop) compressive stress in the plate is of the order of one third of this clamping pressure:

$$\sigma_{hoop}^{compression} = \frac{\sigma_{clamp}^{plate}}{3} = \frac{127}{3} = 42\ MPa \qquad (3.25)$$

The radial compression as well as the hoop tension is certainly of a smaller magnitude than the hoop compression due to the greater stressed volume in the former case (these stresses are self-equilibrating).

In order to prevent slippage in a riveted joint, and consequently to transfer shear load by friction, it is important that not only the clamping force is of a certain magnitude but the friction coefficient for the joined plate surfaces must also be sufficiently high. A coated (i.e. painted) plate surface has a lower frictional resistance than a clean mill scale surface, and it is therefore more susceptible to early slippage at lower load levels. This indicates that riveted joints with coated surfaces have lower fatigue lives than similar joints with uncoated surfaces since they transfer less shear load. The ultimate static strength, however, is not affected – it is still only a function of the shear capacity of the rivet or the bearing capacity of the plate (and the rivet), provided that the connected plates do not fail in tension.

A rivet hole – a severe defect in itself – will become an even more fatigue prone detail, when the hole edges are subjected to corrosion damage.

The common practice in Sweden has been to apply two layers of paint (linseed oil mixed with red lead) to the plate surfaces to be joined. This method may somewhat shorten the fatigue life of the joints, but it has also prevented the initiation of corrosion between the plates. The clamping forces and protective paint help to keep the joint tight and free from corrosion. A rivet hole – a severe defect in itself – will become an even more fatigue prone detail, when the hole edges are subjected to corrosion damage.

The importance of a clamping force, for lengthening fatigue life and preventing initiation of a crack at the rivet hole edges, cannot be stressed enough. The fatigue life of a riveted joint is more dependent on the clamping force magnitude rather than on the rivet configuration or joint detail appearance. It is therefore of vital interest that the magnitude of the clamping force is practically unchanged during the service life of a riveted detail. There are, however, a number of influencing parameters that can reduce the clamping force:

- Fretting of the plate surfaces in contact.
- Relaxation of the rivet and the laterally compressed plates.
- Large shear deformations.
- Yielding (by tension) of the net section.

When two plates are joined together, the surfaces can wear down through the effect of *fretting* if the plates move longitudinally in opposing directions. An adequate clamping force prevents this behaviour during repeated loading since the longitudinal forces are transferred by friction. If the slip load is exceeded, the joined surfaces will move extensively and fretting then takes place. The protective paint will soon vanish and the plate surfaces are then successively worn down. This process results in a loss of plate thickness, which reduces the clamping force little by little. The small particles which

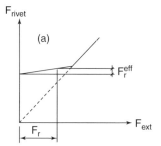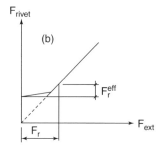

Fig. 3.14 The effective stress range in a rivet subjected to an externally applied cyclic tensile load for (a) large clamping forces and (b) small clamping forces.

are torn away from the surface of the plates form a fine powder of rust which can be seen by the naked eye coming out at the end of the plates or from under the rivet head. This phenomenon is called fretting corrosion. The plate surfaces are also susceptible to fretting fatigue cracks due to local stresses at the surface. These surface cracks act together with any fatigue crack originating from a macro defect (e.g. a rivet hole) and consequently further shorten the fatigue life of the joint.

The *stress relaxation* (i.e. creep-related stress reduction) of the rivet and the compressed plates is a function of stress level, time and temperature. Even though the stress levels are below the elastic limit, irreversible deformations can take place due to dislocation movements. These movements are a function of time and temperature. A rise in temperature over a given period of time will make it more likely that these micro-slip deformations take place. Perhaps the effect of creep in the rivet and the plates is negligible due to the initially low stress levels, but the effect has to be taken into account for other types of pretensioned structures. Prestressed reinforcement bars and strands in concrete structures are expected to lose approximately 10% of their initial tension over a very long period.

Large *shear* deformations induce secondary bending in the rivet shank and this will diminish the clamping force. This is due to the yielding, which plastically deforms the shank. The irreversible deformations will reduce the initial and elastic clamping stress in the rivet, and ultimately this stress will be reduced to zero if the yielding is large enough. This behaviour is the same for rivets in pure tension.

Yielding of the net section by a tensile force reduces the clamping force due to the permanent reduction in plate thickness which takes place.

The importance of a retained clamping force is even more important for a riveted joint in lateral tension. Consider the example in Fig. 3.14, which shows the effective stress range in a rivet for two different initial clamping force levels.

The importance of an adequate and sufficiently high clamping force in relation to the presumed applied service load is clearly shown in Fig. 3.14. If the clamping force is high, it also reduces the possible prying effect (this is discussed further in section 4.2). Fatigue tests have shown that the fatigue life is highly dependent on the prying ratio (i.e. the relation between the prying force and the externally applied tensile force). As the prying ratio is increased, there is a decrease in fatigue life.

Chapter 4

Fatigue of Riveted Railway Bridges

4.1 General

The fatigue resistance of a full-scale riveted railway bridge (or a highway bridge for that matter) is dependent on a large number of different factors. Some of these factors explain the difference in fatigue resistance between a full-scale riveted bridge structure and a small riveted virgin specimen manufactured in a laboratory environment. The additional factors, which have to be considered when estimating the fatigue life of a full-scale riveted bridge structure, are:

– Initial imperfections (i.e. crookedness, inclination, deflection, etc.) in the different elements as well as in the structure itself. Every deviation from the ideal condition induces secondary stresses, which shorten the fatigue life.
– A full-scale riveted structure consists of a large number of identical riveted details that, taken together, have a lower fatigue resistance than a single detail. The "weakest link" will always govern the load-carrying capacity of a "chain", and one can therefore assume a certain "size effect" for large structures.
– Riveted built-up girders show inherent redundant fracture behaviour when experiencing severe fatigue cracking. When a fatigue crack has propagated so that an individual plate or profile in a built-up section has broken completely, the crack is stopped (temporarily) due to the redistribution of forces to undamaged parts. The rivets in the locations near to the damaged section are transferring the forces around the cracked member element. The number of member parts, the more significant is this redundant behaviour.
– Statically determinate structural systems may also exhibit redundant fracture behaviour. The load is gradually transmitted to stiffer parts and elements as a structural member is broken down by fatigue. Consider a simply supported truss where the diagonals are crossed in each section. This type of structure has almost an unchanged load-carrying capacity whether one or two diagonals are carrying the load in a section. Another example of these "internally redundant" structures are short-span bridges, which often consist of four girders laterally joined together with a cross-framing system. If one girder has fractured, the other three are still capable of carrying the load by redistributing forces via these horizontal elements.
– The members of a full-scale riveted bridge are connected to each other by joints of different shapes and sizes, and these differ from the joints normally used in

laboratory fatigue tests. In large riveted joints there are a number of influencing factors that might not be present in a fatigue test of a small riveted detail. Some examples of these factors, which influence the load distribution in the joint, are:

- If the connection length (i.e. total pitch distance) is large, there is a tendency for the outer rivets to carry a greater part of the shear load due to the difference in plate elongation within the joint.
- There can be variation in the clamping force at different rivets. Apart from the effect of the riveting technique itself, this could also be the result of differences in grip length. Loose rivets fall into this category as well.
- A misalignment between two adjacent rivet holes has the effect that some rivets carry a greater portion of shear force by being more in bearing from the beginning than others, see the figure below:

- Forces transverse to the longitudinal direction might be introduced by restricted Poisson contraction.

- Secondary stresses can be present wherever there is some constraint which prohibits bending deformations (i.e. at every joint). These "second order" deformations can be introduced not only during the vertical static deflection of the structure, but also by dynamic vibrations and transverse horizontal forces.
- The presence of different defects, such as corrosion damage, scratches and hit marks, could all result in a shorter fatigue life.
- The stresses due to live load are often difficult to estimate. The dynamic amplification factor could be substantial. Normally though, the observed stresses are below the calculated static value due to some restraining effects (in comparison to the ideal structure assumed in the design process):

  - A truss joint (as well as other types of joints and connections) is often assumed to act as a hinge. Instead these joints have a semi-rigid behaviour in the real structure.
  - There is always some friction in the movable bridge bearings, which restricts the horizontal movement (and consequently the vertical deflection).
  - The presence of horizontal "braking stiffeners" at the ends of a truss bridge makes the stringers act together with the primary truss members, see the figure below:

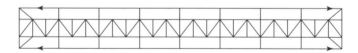

- The steel material properties are not constant throughout the bridge. Profiles of relatively high steel quality are mixed with profiles of lower (but still adequate)

quality. The fatigue life of non-welded details is directly related to the steel quality, and it is therefore quite difficult to predict which one of two details (similar in shape) is more inclined to experience fatigue cracking under the same type of loading. In the past, when most riveted steel bridges were constructed, steel properties had greater variation than is found in contemporary steel. For example, the steel profiles used for a truss bridge were often of rimmed steel. The characteristics of rimmed steel are certainly anisotropic with respect to the chemical content. Carbon, sulphur and phosphorus have a tendency to gather in the parts of a steel profile that experienced the slowest cooling after the steel-making process. Additionally, in these centrally located parts, the grain size is coarse due to the slow cooling process. This results in a fabricated steel material that has somewhat inferior quality (e.g. less ductility) in the central parts of a plate. In addition, rimmed steel has a high oxygen content, which results in a large presence of slag. These impurities give the steel a lower tension capacity, especially in the transverse thickness direction. Taken together, the uneven slag distribution and the difference in ductility throughout the thickness contribute to a great uncertainty concerning the steel properties of any given detail.
- Different riveting techniques are used for different rivet connections. Most of the riveting work is made in the shop in order to prefabricate as much as possible before connecting the different members in the field. The shop practice has been to use hydraulic pressure riveters, which together with the environmental advantages (i.e. with respect to the noise) lead to a satisfactory result (e.g. more consistent clamping forces). In the field conditions are not as favourable, which often leads to unsatisfactory results. Connections can be difficult to access and pneumatic hand hammers are used, which makes the riveting procedure in the field sometimes complicated and hard to carry out. The result is obviously not as good as in the shop, and consequently the load-carrying capacity is also reduced (mainly as a result of a greater variation in clamping force compared with the prefabricated connections).
- If a rivet in a connection becomes loose during its service life (under the influence of a repeated cyclic load), it should, after being detected during the recurrent rivet revisions, be immediately replaced. This can be seen as a successive "up-grading" of the connections and increases their capacity to withstand cyclic loading.

All these factors, which have to be considered when estimating the fatigue life of a riveted full-scale structure, have both an "accelerating" and a "hampering" effect on the initiation of a fatigue crack in comparison with the behaviour of a small riveted detail.

When comparing different full-scale structures, one also finds an additional "accelerating" factor that makes railway bridges more susceptible to fatigue cracking in comparison to highway bridges. A highway bridge has a greater proportion of its load-carrying capacity taken by the dead weight than a similar railway bridge (same span length). The ratio between the stresses induced by the dead weight and the total design load stresses (including the dead weight effect) is of the order of around 50% for highway bridges with a span length in the region of 50 m. The same relationship for railway bridges with similar span lengths is about 15–20%. This illustrates the fact that a railway bridge is generally carrying a greater proportion of its design load by the traffic load and is therefore subjected to a greater variation in live load stresses than a highway bridge. This difference in live load stress ranges between highway bridges and railway bridges, together with the other "accelerating" factors regarding fatigue susceptibility when comparing a full-scale riveted structure with a single riveted detail, suggests that fatigue is a very important and critical issue for a riveted railway bridge. The experience over the years, though, has been:

- That there is a *lower* probability of fatigue cracking in an actual railway bridge than in a highway bridge. This is mainly attributed to differences in joint detailing, together with cautiousness regarding the use of welding as a connection method in railway bridge design procedure. Furthermore, a few decades ago highway bridges were normally not designed with respect to fatigue.
- That the fatigue life of full-scale riveted structures has proved to be *longer* than that predicted by using the results from fatigue tests on small riveted details (see Chapter 2 for the results from different full-scale fatigue tests conducted over the years).

This last remark indicates that the "hampering" effects, previously noted in this section when comparing the fatigue susceptibility of a full-scale riveted structure with a riveted single detail, outweigh the negative influence of the "accelerating" effects. One main influencing factor, which could explain why there are so few fatigue failures in riveted railway bridges, is the absence of stress ranges sufficiently high to initiate a fatigue crack (i.e. the actual stress ranges are below the fatigue limit). In addition to the reasons already given for why the live load stresses often are below the calculated static value, one can add the influence from the prescribed dynamic amplification given in the railway bridge codes (cf. Fig. 1.2). The calculated static stresses should be magnified by a dynamic amplification factor, which can be very high, especially for shorter span lengths. This factor represents an upper limit that takes extreme conditions and probabilities into consideration. Normally though, the actual dynamic affects are of a much smaller magnitude, and the live load stresses therefore have a substantial "safety margin" in relation to the design stress level. The static live load stresses must accordingly be kept sufficiently low in the design process in order to have sufficient margin when taking dynamic amplification into account.

4.2 Fatigue critical details

In an old riveted railway bridge (e.g. a truss bridge) there are a great number of riveted details which are susceptible to fatigue cracking due to the fact that they have been loaded in service for a large number of years. Whether a riveted joint is "fatigue-prone" is governed by factors such as:

- Detailed geometry of the joint.
- Stress range level.
- Number of stress cycles.
- Loading history.

Two identical riveted joints in the same bridge can differ substantially in terms of their susceptibility to fatigue. Even if the stress range level experienced in the joints due to passing trains has been the same, the actual number of stress cycles can vary due to a difference in influence length. If, for example, we compare a truss bar with a stringer, we find these fundamental differences with reference to the number of stress cycles experienced during the passage of a train over the bridge:

- A *truss bar* generally has an influence length equal to the span length of the bridge. In principle therefore, it will only experience one complete stress cycle during the passage of a train.
- A *stringer*, which is "simply" supported at the ends on the floor beams (i.e. the cross beams), will on the other hand experience a substantial number of stress cycles for the same train. This is due to the fact that the stringer has an influence length about equal to the (normally short) distance between the floor beams. The number of stress cycles can therefore often be the same as the number of axles in the train (i.e. two for each carriage).

Consider this simplified example: for a train with 20 heavy carriages (the locomotive included), the number of stress cycles for the maximum stress range is consequently 40 for the stringers in the bridge, while the truss bars only experience one single stress cycle.

If we compare similar riveted members in different bridges, we find an additional factor which, besides the influence length, also contributes to the susceptibility of fatigue cracking. The number of stress cycles experienced is not only dependent on the influence length, but is also a function of the loading history. Two identical riveted members in two different but nominally identical bridges can differ substantially in terms of the fatigue damage accumulated over the years. The fatigue damage accumulation is a function not only of the number of years in service, but of the number of passing trains (especially heavy freight train traffic) during these years. Thus the stress range level, the influence length of the different bridge members (i.e. the number of stress cycles), and the loading history are all very important parameters when deciding upon the potential fatigue damage in an old riveted railway bridge.

There are numerous fatigue critical details and member parts, where one can assume (and also where they, in some cases, also have experienced) fatigue cracking in riveted railway bridges, especially:

- Floor-beam hangers.
- Stringers.
- Floor beams.
- The stringer-to-floor-beam connections.

There have been a number of reported fatigue cracks in floor-beam hangers over the years, especially in the USA (see for example ref. [59]). The fatigue cracks start at any of the four corner rivet holes (there are two joint plates connecting the vertical hanger), see Fig. 4.1.

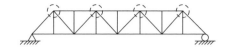

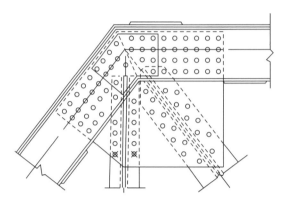

Fig. 4.1 Potential fatigue critical regions in a truss bridge – the connections between vertical floor-beam tension hangers and the horizontal top compression bars [40].

Carter et al. [59] had the working hypothesis that this fatigue cracking was caused by high rivet bearing due to low clamping forces. There is no doubt that a small clamping force is detrimental with respect to fatigue, but suggesting that an assumed absence of clamping force should be the sole reason behind this problem is to overlook the main cause. Due to the deformations in a truss bridge when it is being loaded, the truss joints are subjected to both in-plane and out-of-plane bending. This results in *secondary bending*, which introduces additional stresses in the truss bars, especially in the connection region where there are the greatest constraints against rotation. These secondary bending stresses are often not accounted for in the design process, where instead the truss bars are assumed to be simply connected at the joints (i.e. as free hinges). If we model a truss bridge as a three-dimensional space frame, we normally find that these additional bending stresses are negligible as far as the ultimate static strength is concerned, but in a fatigue design calculation, however, these stresses must be taken into consideration.

It is obvious why the corner rivet locations belong to the critical region in the connection:

– The normal tension force in the vertical bar is not equally distributed among the rivets in the connection. The outer rivets will carry a greater portion of the tension load than the inner rivets due to differences in plate elongation between the rivet locations.
– The secondary bending (both in-plane and out-of-plane) will introduce additional shear forces in the rivets. The corner rivets, which have the largest lever arm, will be the main "resistant factor" against moment rotation for in-plane bending.

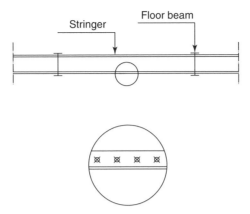

Fig. 4.2 Potential fatigue critical region in a stringer – the lower tension flange rivets at midspan.

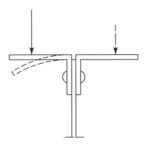

Fig. 4.3 Potential fatigue critical member part in a stringer – the outstanding leg of the top compression flange.

For out-of-plane bending (due to bending of the stringers), the shear force is unevenly distributed for the same reason that explains the uneven distribution of a normal tension force.

For a longitudinal stringer, one can assume that potential fatigue damage has accumulated in the midspan lower tension flange, see Fig. 4.2.

The stringers are normally short (say 4–6 metres) and are therefore subjected to a large number of stress cycles for each passing train. The lower tension flange at the midspan is subjected to the largest stress range, and the rivet holes are the most severe defect in this region. If the tension flange has an additional flange plate connected by rivets to the L-profiles (as shown in Fig. 4.2.), these new rivets will further increase the possibility of fatigue cracking. Often these stringers have a diagonal horizontal bar connected to the lower flange. This connection can, if it is located at or near midspan, also govern the fatigue life of the stringer.

If the sleepers are directly placed on top of the stringers (which is the normal procedure), there is a possibility of fatigue cracking in the outstanding leg of one of the top L-profiles due to eccentric loading, see Fig. 4.3.

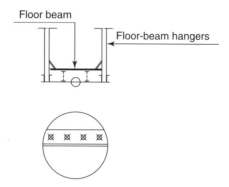

Fig. 4.4 Potential fatigue critical region in a floor beam – the lower tension flange rivets at the midspan between the stringer connections.

When the outstanding leg is subjected to eccentric loading from the sleepers it will bend, and this might therefore induce a longitudinal fatigue crack at the corner of the L-profile. The rivets at these locations are also fatigue critical. The shear stress range in the rivet shank is a function of the combined effect of the horizontal shear forces due to composite action and the shear forces due to the vertical loading. In addition, there are also some bending stresses due to the flexure of the L-profile. If prying of the connection is taking place, there is also an additional normal tension force to be added to the bending stress.

The floor beams are loaded vertically at the two locations in the web plate where the longitudinal stringers are connected. Since the stringers are equally loaded, the floor beams have a constant-moment region at the midspan between the stringer connections. In this region, the lower tension flange rivets are equally critical with respect to fatigue. The rivet hole at any of these locations is the inevitable stress raiser that sooner or later might initiate a propagating fatigue crack, see Fig. 4.4.

There is no evidence that the moment-resistant connections between the floor beam and the vertical hangers – see Fig. 4.4 – are fatigue critical, even though one can assume a constantly repeated heavy loading from trains. The connection is mainly transferring the vertical loading from the stringers to the truss hangers by shear forces. Although the connection is very stiff, only a small moment is transferred due to the high flexibility of the vertical hangers (in comparison to the floor beam). The diagonal plate profiles at the ends are, besides helping to transfer the vertical load, also acting as reinforcement for the hangers. The bending moment in this connection is negligible in comparison to the maximum midspan moment for the floor beam, but for the vertical hangers the same loading is producing a substantial bending stress in this region, which cannot be neglected.

The connection between a stringer and the floor beam is perhaps the most difficult non-moment resistant joint to design for fatigue. In order to achieve a sufficient shear load transfer capacity in the connection between the stringer and the floor beam, it is unavoidable that the connection will have a certain flexural stiffness, which resists some bending moment as well. This behaviour is normally minimized by only connecting

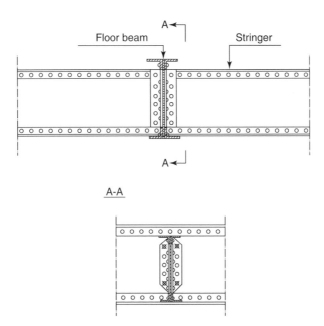

Fig. 4.5 Potential fatigue critical rivets in the connection between a stringer and a floor beam. This particular stringer-to-floor-beam connection is identical to the one used in the railway bridge over the Vindelälven at Vännäsby (for further information about this bridge, see Chapter 6).

the stringer web plate by angles to the web of the floor beam. Over the years, however, there have been a number of fatigue failures in these connections, where the heads of the top and bottom rivets in the outstanding legs have fallen off, see Fig. 4.5.

This type of fatigue failure in the rivets is explained by the additional normal forces, which – besides the unavoidable bending moment in the connection – is also a function of:

- Possible prying.
- Localized bending of the rivet head when the connection is subjected to shear deformations, as shown in the figure below:

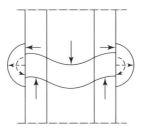

Fig. 4.6 Fatigue failure of a rivet in the connection between a stringer and a floor beam in the railway bridge over the Vindelälven at Vännäsby. The failed rivet is the critical one shown top left in Fig. 4.5.

Eventually, after being subjected to a sufficient number of repeated loading cycles, the rivet head will fall off due to fatigue in the rivet shank. Fig. 4.6 shows a rivet which has failed in the web angle connection between a stringer and a floor beam.

For the bridge at Vännäsby, the normal procedure over the years has been to replace these failing rivets one by one with high-strength friction-grip bolts.

In a way, these failures are not as critical for the load-carrying capacity as one might think. The main purpose of these connections is to carry the shear load and transfer it to the floor beams. If a rivet head drops off due to repeated bending, the connection will only lose some of its rigidity against moment rotation (which was the reason for the local rivet head failure). The shear capacity will be largely unaffected as long as that the rivet shank is intact and still in its place.

It is not easy to foresee when and where a rivet will fail in these kinds of connections. First of all, it is difficult to estimate the magnitude of the bending moment due to traffic loading in these semi-rigid connections. These factors all contribute to a large uncertainty concerning the magnitude of the bending moment in these stringer-end connections:

- Flexibility of the floor beams.
- Flexibility of the web angles.
- Elongation of the rivets in the outstanding angle leg.

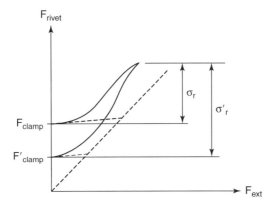

Fig. 4.7 When the initial clamping force is reduced by plastic deformations and fretting, there is an increase in the effective normal stress range level in a rivet for an externally applied tension load. The resulting load characteristics for the rivets given in this figure are for a case when prying can be considered an influencing factor (full curves). As can be seen, the ultimate static strength is independent of the clamping force level. The dashed lines show the case without prying, cf. Fig. 3.9.

- Shear deformations of the rivets.
- Bearing deformations.
- Friction slip in the riveted joints.

The lifespan of this kind of connection is a function of the loading level, but it is also highly dependent on the clamping force. The riveted joints between a stringer and a floor beam are normally executed in the field, and they are therefore not fabricated as well as those made in factory work shops. The clamping forces in these joints can thus be expected to vary to a greater extent than those in the "prefabricated" joints in the bridge. In addition to the uncertainty concerning the magnitude of the initial clamping forces in the joint, it is also very difficult to predict the level of the normal stress range in the different rivets. Although the resulting bending moment in the semi-rigid connection can perhaps the predicted (taking different flexibilities into consideration), there will still be some difficulties in estimating the effective stress range level valid for loading over a long period of time. As the clamping force is gradually reduced through fretting and bending (prying and localized bending of the rivet head are the important parameters here), the normal stress range level in the rivets will increase. Consider the example given in Fig. 4.7 for two different clamping forces.

It is easy to see, though difficult to calculate, why the fatigue life of the top rivet in a typical stringer-to-floor-beam connection is reduced compared with a rivet under pure shear. The negative bending moment introduces axial tensile forces in the rivets at the upper part of the connection. But why do the bottom rivets also fail by fatigue in these connections (cf. Fig. 4.5)? The explanation is that they experience a similar tension stress range level as the top rivets. The cyclic compression in the lower parts of the connection causes a cyclic change in the tension in the rivets due to the initial clamping force. As is also the case with the top rivets, the bottom rivets carry a greater part of the shear load than the inner rivets. One can therefore assume a larger influence from localized bending of the rivet head for these outer rivets. For the railway bridge over the Vindelälven at Vännäsby, the rivets themselves have been the "critical spots" in the connection between the stringer and the floor beam. This might partly be due to the fact that the bridge was built in 1896, and there is therefore reason to believe that the riveting was carried out with hand hammers. The actual outcome can then be assumed to be resulting from rather poor quality rivets with a great variation in the initial clamping force in each joint. For connections of the same type and detailing, but for bridges built in the twentieth century, the rivets can be assumed to be a lot more resistant against fatigue failure. Instead, the angles in these later bridges can be the "weakest point". One should therefore also be observant to possible cracks at the angle corners (the top and bottom parts).

All the members and details that have been described in this section as being critical with regard to potential fatigue cracking have one important parameter in common. They all have short influence lengths. (This was noted as an important parameter at the beginning of this section.) Wherever there is a bridge member with a short influence length, one can assume a large number of stress cycles for each passing train. More attention should therefore, as a rule, be paid to these members than to the other members in the bridge (e.g. the different truss bars). However, one should be aware of the fact that rivets may work loose in any riveted connection, and if they are not immediately replaced, the fatigue strength is adversely affected. Every riveted joint designed to only transfer normal forces induces secondary restraint to bending of the second order (in-plane and out-of-plane bending), and they should therefore be checked from time to time regardless of whether the joints are assumed to be fatigue critical. But in a way, if a rivet works loose in a joint due to secondary restraints, and the joint then becomes less rigid, this in many cases can be favourable. It is not so obvious to decide whether a loose rivet is in a fatigue critical location. As was noted at the beginning of this section, stress range and the number of stress cycles are the governing parameters, and if these are low in a joint, the "tolerance" for secondary stresses are high.

4.3 Dynamic amplification

The dynamic amplification of the static response of a railway bridge subjected to train loading is mainly a function of different mass inertia forces. The train load is decelerated into the structure by track and train irregularities, and also by the vertical deflection of the bridge itself. Every surface roughness that deviates from the ideal smooth and plane condition induces small impacts (read, mass inertia forces from the train load), which magnify the static response (i.e. the strains in the structure from the same train standing still on the bridge). These impacts can theoretically induce resonant vibration in the bridge structure, but usually the energy does not last long enough nor are the bridge members easy to excite due to the (normally) large masses involved. In addition, the bending stiffness of the bridge members is high, which results in high natural eigenfrequencies (especially for stringers and floor beams). A high natural frequency for a bridge element means high structural damping (i.e. energy losses during deformation). For a riveted bridge member, one can assume a substantial amount of inner damping coming from the friction slip of the connected plates (a kind of damping not to be found in welded or rolled profiles). The bending stiffness of the members in an old riveted railway bridge is also high in comparison to a newly designed bridge. The low steel quality and lower design stress levels in the old riveted railway bridges result in stiffer members in general, which give them higher eigenfrequencies.

The only type of riveted railway bridge structures that might be susceptible to resonant vibration are lattice bridge girders (i.e. where the girder web consists of flat plate diagonals which cross each other). These flat diagonals have a low natural eigenfrequency, and a small dead mass, which make them very easy to excite by the vibrations from the train. However, lattice bridge girders are no longer common (and are certainly rare in Sweden), and they are perhaps not of great concern.

The dynamic amplification related to the vertical deflection of the bridge itself is, for normal riveted bridges, the result of a dynamic displacement which is forced upon the bridge, not an interaction of the train load frequency and the eigenfrequency(-ies) of the bridge structure (read, resonant vibration). To excite resonance between the train loading and, for example, a simply supported riveted girder bridge, the loading recurrence time (related to train velocity) must be about the same as the vibration period for the first eigenmode of the bridge. The bending stiffness of riveted girder bridges in general is of such a magnitude that the lowest eigenfrequency (for bridge plus mass of the train) is sufficiently high to exclude the possibility of resonance. The "critical velocity" for these bridges can be estimated at 500 km/h or more.

The passing trains are anyhow inducing dynamic forces, with or without initiating resonant vibration. When a bridge structure deflects under load, mass inertia forces are being induced in the structure due to the displacement. The train load is decelerated vertically by the flexible bridge structure, and this amplifies the static response. From the derivation of centrifugal forces we find the following relationships, which are also applicable to the dynamic amplification of the static response for a train that passes over a (flexible) bridge:

- As the radius of curvature (which depends on the span length) decreases, the centrifugal forces (i.e. the mass inertia forces) increase.
- As the tangential velocity (i.e. the train speed) increases, the centrifugal forces also increase.

For a given span length, the radius of curvature (read, deflection) is also dependent on the train load. As the train load increases, the deflection increases, and consequently so do the mass inertia forces. The heavier the train, the more dynamic amplification is to be expected in any given bridge.

The dynamic amplification factor given in the railway codes (cf. Fig. 1.2) can be seen as the "maximum combined effect" of all the dynamic forces that could be assumed to act on a bridge structure. This factor should not be taken as the dynamic amplification that occurs from normal "everyday traffic"; instead it should be regarded as a "safety factor", which only should be used in the design process. For everyday train passages the amplifying effect on the static response due to dynamic forces is often "counter-balanced" by stiffnesses in the structure that are not accounted for in the design process (e.g. semi-rigid connections assumed to be acting as hinges).

In conclusion, the general advice to be given if one wants to make a fatigue damage accumulation analysis of a riveted railway bridge (i.e. an estimation of the stress ranges for everyday traffic) is to ignore the dynamic forces and just consider the static response. In the structural design codes the specifications on this matter are rather ambiguous (or, simply speaking, such considerations are omitted). However, if one can assume excessive deformations due to resonant vibration in a structural member, it goes without saying that this should be duly taken into consideration. A structural member or detail that is subjected to resonant vibration will, of course, normally have a heavily reduced fatigue life.

4.4 Brittle fracture

Brittle fracture can be characterized as a failure mode which occurs without any prior warning and without excessive deformations due to yielding. The nominal stress could be well below the yield strength, but due to a stress raiser, in combination with several other factors that make the material more brittle, the material can experience so-called cleavage fracture. These additional factors, which enable the stresses to remain elastic at a localized strained point, are:

- Temperature – when the temperature is lowered the distance between the atoms is decreased, and therefore the shearing deformations, due to slippage of the dislocations, are more likely to be prevented.
- Loading rate – the faster the strains and stresses are being transferred into the material, the more the movements of the dislocations will be prohibited. A dislocation movement (and consequently a shear fracture) always requires a certain amount of time to develop.
- Material properties – if the quality of a material (e.g. steel) is improved (i.e. the tensile strength is increased), the ductility is normally lowered. The perlite grain content is greater in a steel of high quality than in a steel of an inferior quality (the carbon content increases at the same rate as the presence of perlite grains increases). A steel can have a low ductility, despite its quality characteristics, due to other features of its chemical content and metallurgical factors:

  • Sulphur and phosphorus make steel less ductile.
  • Nitrogen can diffuse out of the steel and, as it does so, attach to the dislocations, which "locks" them to the rest of the ferrite atom structure (this phenomenon is called ageing).
  • Rimmed steel has a high content of different oxidation products, which makes these kinds of steel more brittle than other steels (read, semikilled or killed steels, where the oxidation products have largely been removed), especially in the transverse thickness direction.
  • Due to slow cooling during the steel fabrication process, large grains can develop. As the grain size increases the ductility is decreases.

- Plate thickness – as the thickness of a steel plate is increased, the susceptibility to brittle fracture is also increased. When yielding is taking place due to large strain deformations at a notch (read, stress raiser), the surrounding material will deform transversely to the loading direction due to Poisson type deformations. The surrounding material, still under elastic strain, will prohibit these transverse strains in the plate width direction, and therefore introduce additional (tensile) stresses at the notch. And when the plate thickness is increased, these "strain restricting stresses" will also occur in the thickness direction, therefore subjecting the notch to a triaxial stress state.

In the 1930s, when the welding technique was being introduced into railway bridge construction (enabling parts and members to be joined together more rapidly than before), there were some catastrophic failures in Belgium and Germany that could be attributed to brittle fractures. The reasons why these failures happened can be summarized as follows (they can be seen as brittle fracture cases where all the important influencing parameters were present):

– When it was first introduced, the welding technique was not fully developed (and far from perfected). Poor detailing in combination with improperly performed welding created stress concentrations that were definitely not considered in the design process.
– In many cases the steel was too brittle (often because the carbon and/or nitrogen content was too high) to be suitable for welding.
– As well as using a brittle material, high-hydrogen electrodes were being used, which made the material even more brittle after welding.
– Instead of joining several (thin) element parts together as had been done in the "riveting era", very thick plates (up to 60 mm) were welded together without prior pre-heating. The residual tensile stresses in the welded joint therefore became high, in combination with excessive grain sizes due to the large heat input.
– The welding procedure started at one end and finished at the opposite end of the structure. This creates an unnecessary high presence of residual tensile stresses in the welded region due to the thermal constraints introduced.
– There was at the time very poor knowledge about the fatigue resistance of a welded joint.

All these factors, in combination with low temperatures and a dynamically loaded structure (e.g. a railway bridge under train loading), contributed to the initiation of a brittle fracture, which in some cases led to a complete and catastrophic failure.

These examples of brittle fracture represent failures because the toughness of the steel (and the structure) was of insufficient magnitude. The initial flaw (e.g. a crack) very soon became sufficiently severe (e.g. a longer crack) to initiate a brittle fracture. The fracture toughness of the material was sufficient to withstand stable crack growth, instead the crack propagation became unstable (i.e. brittle fracture was then taking place). The "crack growth history" of these bridges can be summarized as follows:

1. Initial large cracks due to improper welding of the brittle material have to be assumed from the start (i.e. from the first day the bridge was taken into service).
2. These initial cracks then propagated (more or less rapidly) until at least one reached the so-called critical crack length.
3. A brittle fracture was then initiated.

One could say that the "initiation phase" (i.e. the time required to initiate a small micro-crack near or at a notch) had been concluded at the start of the bridge's working life. Normally, in non-welded joints such as riveted connections, this initiation phase covers the major part of the total fatigue life. This difference between fatigue/fracture behaviour in a riveted connection and a welded joint illustrates the different phases in crack propagation:

- Phase 1: A certain number of stress cycles of a particular magnitude is required to initiate a small micro-crack near or at a notch.
- Phase 2: This micro-crack then propagates during continued cyclic loading.
- Phase 3: The final failure is then either brittle or ductile due to factors such as:

  – Stress level.
  – Stress distribution and stress gradient.
  – Loading rate.
  – The ability of the material to redistribute stresses by plastic flow (i.e. fracture toughness).
  – Structural redundancy (the ability of the structure to redistribute forces to stiffer regions).
  – The ductility of the material.

Research concerning the phenomena of fatigue and fatigue life is dealing with questions like:

– Why does a crack initiate at a notch?
– How many stress cycles are required to initiate a crack, and then how many are needed to propagate it until final failure of the section?

Research on fracture mechanics is merely concentrating on the last two of the three phases (mentioned above) or to questions such as:

– How fast does a crack propagate during different stages?
– What is the size of the defect (read, crack) when unstable crack growth (i.e. brittle fracture) takes place?

The initiation of a fatigue crack at or near a rivet hole in a riveted connection "consumes" approximately 90–95% of the total fatigue life. Parameters such as stress level, loading rate, material characteristics and temperature all influence crack propagation after crack initiation, and they also determine whether the final failure of the net section is ductile or brittle. The fracture toughness of a material is highly dependent on the temperature, and therefore the final stages of a cracked element are also a function of this parameter. At low temperatures, the final failure will start sooner and will predominantly be brittle (the critical crack length is shorter). If the final 5–10% of the total fatigue life of a riveted connection is a function of temperature, is the "initiation time" also dependent on temperature? The fact is that when the temperature falls, the material contracts and it becomes harder, and so the fatigue limit σ_{ru} is actually increased, see Fig. 4.8.

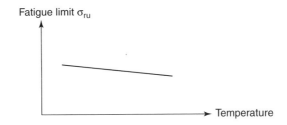

Fig. 4.8 Principal correlation between fatigue limit σ_{ru} and temperature.

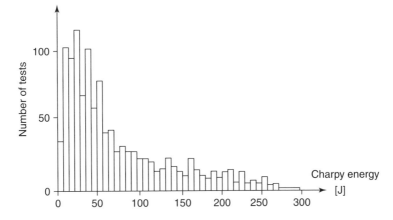

Fig. 4.9 A summary of over 1200 Charpy V-notch tests at 5°C (40°F) of steel samples taken from old riveted bridges [24].

When the temperature falls the distance between the atoms decreases consequently preventing movements of the dislocations. The material becomes harder (reduced ability of plastic flow) and the initiation of a crack will be characterized by separation at the grain boundaries rather than by the formation of slip bands (i.e. cleavage fracture instead of shear fracture).

The steel material from old riveted railway bridges has been shown to be very brittle according to results from impact toughness testing. Fisher et al. [24] collected data from over 1200 Charpy V-notch tests performed on steel samples taken from about 90 riveted bridges built between 1890 and 1955 in the USA, see Fig. 4.9.

In the Fig. 4.9, it can be seen that many of the test results show impact energy values of 50 Joule (Nm) or less. In Sweden, where the temperature can be as low as −30°C (sometimes even lower) especially in the northern parts of the country, these results suggest that many old riveted railway bridges will not satisfy the structural codes on impact toughness. A minimum of Charpy V-notch impact toughness measure of 27 J is required at the lowest temperature to be expected in service (or, in other words, the transition temperature − i.e. when the fracture behaviour changes from ductile to brittle, cf. Fig. 4.12 − should be below the lowest temperature likely to be experienced in service). Kjell Eriksson [36] has given an account of the Charpy V-notch tests that have been performed by the Swedish National Rail Administration over the years. The steel samples have been taken from damaged structural elements in old steel railway bridges (welded and rolled as well as riveted). The impact toughness at −30°C has in general been around 5 J, which is far below the requirements given in the codes.

There are several reasons why this inferior impact toughness of old steel railway bridges has not resulted in any serious fracture failures:

- The specifications on impact toughness emanate from the experience of the Liberty ship disasters, and are therefore, in fact, only applicable to welded plates of thick dimensions. However, the built-up plate elements in a riveted railway bridge have:

  • Plates of relatively thin dimensions (say 8–12 mm).
  • Relatively small residual tensile stresses.
  • An inherent structural redundancy (several parts are joined together to achieve composite action, but they still, in a way, can act separately).

- Field tests and investigations of riveted railway bridges have shown that:

  • Live load stresses are in general relatively low.
  • The strain rate is relatively low as well.
  • The probability of the occurrence of larger defects is small (the live load stresses have been below the fatigue limit, and hence no fatigue crack has been initiated).

This question about impact toughness of railway bridge steels could be transformed into this relevant question: what is the largest defect to be allowed? The author is convinced that if some fracture toughness tests on old riveted built-up thin section elements were to be performed (the author is not aware of any such tests – a reason might be that impact toughness tests and fracture mechanics investigations were a relatively new research areas in the early 1990s (when this thesis was written) and interest had at the time therefore mainly been focused on welded steel elements), taking the actual expected maximum stress level and strain rate into consideration, the toughness of the steel will be adequate to withstand very large defects (such as long fatigue cracks or quite severe hit marks), even at low temperatures.

One key parameter is the plate thickness, which as noted above is one reason why one can assume fracture-tough steel in riveted railway bridges. Consider the illustrative example given in Fig. 4.10. This is a "correction diagram" from British Standard (BS 4741) showing the correlation between Charpy V transition temperature (at 27 J) and the lowest allowable service temperature taking the actual plate thickness dimensions into consideration.

Fig. 4.10 clearly indicates that even though the specifications given in the codes regarding impact toughness are not fulfilled (i.e. at least 27 J at the lowest in-service temperature), the steels from old riveted railway bridges are sufficiently ductile due to the fact that the plate thicknesses in these bridges are generally relatively small. Remember that the Charpy impact testing method was developed to meet the need for a proper procedure to determine the fracture toughness of welded steels of thick

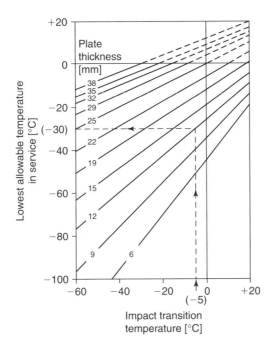

Fig. 4.10 The correlation between Charpy V transition temperature and the lowest allowable in-service temperature according to BS 4741. Taking an impact transition temperature of −5°C as an example (a fairly typical result when performing Charpy V-notch impact tests on steels from old riveted railway bridges), we can see that the lowest allowable temperature for a maximum plate thickness of 12 mm (thicker dimensions than that are unlikely to be found in an old riveted railway bridge) will approximately be −30°C.

dimensions. But why is a thin plate more ductile (read, fracture tough) than a thick plate? The reasons for this are that:

- There is a smaller probability of triaxial stress conditions at a notch in a thin plate. The strain constraints are higher in a thick plate.
- A thin plate generally has small grain sizes due to a fast cooling process (large slip bands are prevented from developing – they are stopped at an early stage by the grain boundaries – and therefore the development of large macro-cracks are hampered).
- The fast cooling process also makes the steel in a thin plate more homogeneous and isotropic. A thick steel plate, which cools slowly after rolling, has a tendency for an uneven distribution of the chemical components throughout the thickness. In particular, carbon and oxidation products can gather in the central parts (which have cooled the slowest), and this therefore reduces the ductility in the transverse (thickness) direction.
- There is a definite "size effect" in thicker plates. The probability of a certain defect increases as the plate thickness (volume) increases.
- The temperature gradient during the cooling process of a thick plate is greater than that in a thin plate. This is the reason why thick rolled plates have higher residual stresses.

A thin plate can become more brittle after *welding*. This is due to the fact that:

- Through the shock treatment, the material in and around the weld becomes embrittled.
- Subsequent ageing makes the steel even more brittle with time.
- The welded joint is normally a large macro defect (i.e. a stress concentration point where the nominal stress is raised).
- Small micro-cracks together with impurities (always present after finished welding) may increase the local stress even more than the macro defects.
- There is a presence of a complex, triaxial residual tensile stress state at the weld.
- A welded joint in a connection is often very rigid and non-flexible (which makes it more difficult for tension relief to take place).

As has been stated above, a riveted built-up section joint can be expected to have substantial fracture toughness compared with a similar welded joint of a single (thick) plate. The specifications regarding impact ductility according to Charpy testing need not be applied to old riveted steel plates. The fact that no serious failures have occurred over the years just proves the "bluntness" of the testing method and the associated specifications.

Fracture mechanics testing has proved to be a more appropriate method to determine the fracture toughness of a material. These tests make it possible to take the stress level, loading rate, size of defect and thickness into account. All these parameters are of upmost importance, and it is possible to achieve results that more realistically relate to the actual in-service conditions of a structure.

One main parameter derived by fracture mechanics testing is the fracture toughness K_c (in non-linear analysis, J_c). This parameter is a constant material property, and it should not be mistaken for the toughness of a material/structural detail to withstand fracture depending on parameters such as defect size, stress level, loading rate and so on. The fracture toughness K_c of a material is defined as the critical value of the stress intensity factor K. The stress intensity factor is a function of nominal stress, crack length and geometry, and it describes the "stress intensity" at the crack tip. When the stress intensity factor reaches a critical value (i.e. K_c), unstable crack growth is taking place (i.e. brittle fracture). Although the fracture toughness K_c is supposed to be constant, the theory is highly inadequate. The fact that the fracture toughness K_c varies with such a vital parameter as the loading rate has led the author to believe that we have to reformulate the theory. Therefore a new parameter called "critical strain rate" is proposed here (within the framework of fracture mechanics).

Critical strain rate – a new design concept?

We start by quoting Rolfe et al. [41]:

> "The effect of temperature is well known and has led to the transition-temperature approach in designing to prevent fracture. However, the effect of loading rate may be equally as important, not only in designing to prevent fracture, but in understanding the satisfactory behavior of many existing structures built from structural materials that have low impact toughness values at their service temperatures."

The strain rate or the loading rate is, in the author's opinion, absolutely the most important parameter to consider when describing the nature of unstable crack propagation. As the strain rate increases the redistribution of forces (read, stresses) by plastic flow at a notch is reduced. The elastic strain energy at such points then increases and it eventually reaches a critical value, which initiates cleavage fracture. Shear fracture (through the slippage of dislocations) is gradually lost as the strain rate increases. At a certain level the yield point and tensile strength will be the same, and the material will break without prior yielding (i.e. experience brittle fracture at small deformations).

But how can one use "critical strain rate" as a design tool? By calculating the maximum values of the stress level and the loading rate, it should be possible to determine the largest allowable defect (e.g. crack length) in a structure such as a railway bridge. Standardized tests should be performed in advance to ascertain the critical strain rate at a notch (calculated as the enlarged nominal strain rate at this particular notch), i.e. the strain rate that will initiate cleavage fracture. This should be done under fixed values of stress level, plate thickness, steel quality and temperature. Tests should be made for different crack lengths in order to establish the critical strain rate for various sizes of defect. The main difference here, in comparison to ordinary strain rate mechanics testing, is that the "critical strain rate" parameter varies with the severity of the defect, while the fracture toughness parameter K_c does not. The critical strain rate will therefore reflect the true behaviour, which makes it easier for civil engineers (and others) to understand the phenomenon.

Ageing
Another phenomenon that is also not fully understood is the ageing of structural steel. Metallurgists may be aware of this phenomenon, especially for cold-formed steels, but how can engineers take ageing into consideration when designing, for example, new bridges or when analyzing existing ones? First of all, the phenomenon of ageing must be defined correctly (and not confused with some general "destruction" of the material over time):

> Ageing is the successive change in the material properties that takes place especially after a cold-working process. The steel becomes hardened (the yield point and tensile strength are increased) and embrittled (the ductility is decreased). A cold-working process produces an enormous amount of dislocation. These dislocations constitute potential "congregation points" for, primarily free nitrogen atoms. By the process of diffusion (the pressure levelling of different gases), these nitrogen atoms can successively "lock" to the ends of the dislocations where there is a free space in the lattice structure. The diffusion rate is a function of time and temperature, and consequently these are the two main influencing parameters (in combination with the nitrogen content) governing the ageing effect.

The principal effect of ageing on the tensile-strain relationship is shown in Fig. 4.11.

(a) After unloading and subsequent immediate reloading, the load-deformation characteristics follow the same curve as the original "working curve". Due to strain hardening effects (the increase in number of dislocations), the yield stress (proof stress) will increase.

66 Fatigue Life of Riveted Steel Bridges

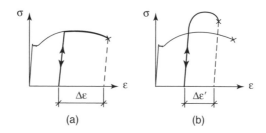

Fig. 4.11 The tensile-strain relationship after an initial preload.

(b) If reloading occurs after a long period of time, then an effect of ageing may be present. The yield point and the tensile strength are both increased, while the ductility, on the other hand, is reduced.

Ageing is to be expected after a cold-working process, but how can such an "overload" become present in, for example, a riveted steel railway bridge? There are several occasions or situations, when there might be reasons to presume a locally concentrated "over-straining". Examples are:

- When the riveting process produces some residual (tensile) stresses due to punching of the rivet holes, or due to localized yielding around the rivet holes from the hammering.
- When the concentration of the normal stresses around the rivet hole is enough to produce local yielding (applicable to a riveted joint where there are no shearing forces to be transferred).
- When the bearing pressure (rivet shank against the plate hole edge) is high (applicable to a riveted joint where there is a significant shearing force to be transferred, e.g. a locally concentrated vertical load of the top flange of a built-up I-section).
- Localized excessive yielding around the tip of a large fatigue crack.
- Cold deformation work due to collision damage.

For a railway bridge, the first four items on this list are normally negligible with respect to a cold-working process leading to subsequent ageing. The last item though – different collision damage – is perhaps when ageing has to be taken into account. Collisions with the superstructure by passing trains (perhaps because goods are sticking out) or by vehicles passing under the bridge will normally produce large plastic deformations. If the bridge member which is hit is then subjected to a large number of (tensile) stress reversals, there is reason to assume that a fracture will occur much sooner than predicted by the nominal fatigue life. This "fracture process" can be subdivided into four stages:

1. The collision damage not only produces a large cold-deformed steel volume, but also a most severe defect, which acts as a large stress raiser.

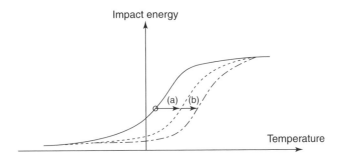

Fig. 4.12 The effect on the transition temperature (from ductile to brittle fracture behaviour) by (a) cold-deformation work and (b) ageing.

2. The material around the damage is subjected to an initial decrease in its ductility due to the yielding taking place, and this ductility is even more reduced if the steel is then subjected to ageing (through a presence of nitrogen in combination with time and temperature), cf. Fig. 4.11.
3. A fatigue crack is subsequently initiated in this area, through a combination of the nominal cyclic stress level and the "notch effect" from the stress raiser (in point 1).
4. Due to the embrittlement of the steel (the yielding in combination with subsequent ageing), the fatigue crack will very soon become "critical" and, if the temperature is sufficiently low, the initiation of a brittle fracture will occur.

The final fracture will either be brittle or ductile depending on the temperature. The probability of a brittle fracture is markedly increased after a cold-working process and subsequent ageing. In a Charpy impact test, the transition temperature is higher when the steel has been subjected to prior cold-forming and ageing, see Fig. 4.12.

Ageing has a "diminishing" effect on the fracture ductility while, on the other hand, it improves the fatigue resistance as well as the static load-carrying capacity. When the yield point and the tensile strength are raised, the fatigue limit is also raised (i.e. it takes a higher stress range to initiate a fatigue crack). However, the critical crack length (with reference to unstable crack propagation, i.e. brittle fracture) will decrease, and this will shorten the fatigue life once a crack has been initiated. As has been stated before, 90–95% of the total fatigue life of a riveted joint consists of the crack initiation phase, and therefore a reduction in the crack propagation time will have an almost negligible effect on overall fatigue life (note that the crack propagation *rate* is unaffected). It is only when there is localized collision damage or similar events, which creates a sharp notch in combination with cold deformation work, that ageing has to be taken into account in lifespan calculations.

Some steels are more susceptible to ageing than others. It is well known that the Thomas process (a special kind of the Bessemer process) leads to a steel type which is more inclined to ageing than other steels. The high phosphorus and nitrogen content makes Thomas steel more brittle and more susceptible to the effect of ageing after a cold deformation process. At the refining process (i.e. when the carbon content is being reduced through adding oxygen to the melted steel) ordinary air is used in the Thomas process, which brings nitrogen as well as oxygen into the steel. The nitrogen that is not solved in the steel (read, into the iron lattice structure) is free to diffuse out of the steel (with the "help" of time and temperature) and, in doing so, locks to the available free atom spaces at the dislocation ends.

When manufacturing new steel, the process being used is obviously well known to all parties involved. But how can one distinguish a Thomas steel from other steel types when investigating an old riveted railway bridge? The chemical content must, of course, be analyzed in a laboratory (using small steel chips taken from the bridge), and then a combination of high phosphorus content (say 0.050% or more) together with a high nitrogen (say 0.010–0.015%) will "reveal" a Thomas steel. Note that Thomas steel normally has a nitrogen content that is at least twice as high as other steels that also have a high phosphorus content. Some typical values of the nitrogen content in different steel types are:

Thomas 0.012%
Martin 0.005%
LD 0.004%
Kaldo 0.003%

If Thomas steel has been killed (i.e. one has practically removed different impurities, such as oxidation products, by adding aluminium or silicon) the ageing tendency is markedly lowered. Nitrogen has a tendency to attach to these additives, especially to aluminium. Naturally, in a semi-killed steel the ageing "resistance" can be expected to be somewhere between that of a rimmed and a killed steel.

So, if the nitrogen content is high, it does not always mean that the steel is susceptible to ageing. As long as the aluminium content is high, and one therefore can assume a killed steel, the ageing effect can be expected to be low (it will then be a function of the carbon content rather than of the nitrogen content). There is little possibility of finding large amounts of aluminium in steel manufactured before, say, the 1930s. All steels manufactured before that period can be assumed to be rimmed (and consequently have low aluminium content) due to the fact that the process of "killing" steels had not been introduced.

The main question is how to identify those bridges that are susceptible to ageing. One has to scrutinize the bridge stock and look for bridges that are prone to being hit by passing vehicles (including those passing on top as well as under the superstructure). Then a thorough investigation of the chemical content of the steel should be performed, and if steels of the Thomas type are found, the bridges should then be kept under special surveillance.

4.5 Corrosion

Corrosion is by definition the oxidation process that unprotected steels (and other metals) are subjected to by the presence of water and oxygen. The corrosion rate is also influenced by temperature and chlorides. If the moisture content is below 60% or the temperature below 0°C, the corrosion process in steel is normally negligible. However, the air moisture content almost always exceeds 60% in outdoor climate, and therefore structures such as railway bridges must be covered by rust-protective paint. A railway bridge is generally protected against corrosion by two different layers of paint. First, a main protective layer is applied on the steel surface. For riveted railway bridges, the first layer normally consists of a linseed oil paint with a lead additive. This layer serves as the real protection against corrosion. The second layer, which covers the first layer, is only there to protect the first layer from destructive environmental forces (e.g. wear and ultraviolet rays). However, if maintenance is poor, the protective layers of paint can be worn down with time, and the exposed steel surface can gradually begin to corrode.

There are several different types of corrosion that might be found in a (riveted) railway bridge:

- An evenly distributed surface corrosion. A general loss of surface material results from this type of corrosion. On the coast or in industrial regions of Sweden the corrosion rate can be 0.5 mm every ten years, but for the inland parts of northern Sweden the corrosion rate can be expected to be a tenth of this value at most.
- Pitting corrosion. This type is concentrated on very small isolated areas, and the corrosion rate can be considerably higher than the rate of general surface corrosion.
- Crevice corrosion. This can occur wherever there is a small confined space (e.g. between two interfacing plates in a riveted joint) that can be filled with water. The inner parts, where the oxygen content is low, serves as an anode, and the electrons flow to the outer parts (read, cathode) which are rich in oxygen.

The first corrosion type (evenly distributed surface corrosion) can be seen as *global* damage, which gradually increases the nominal stress level due to the loss of area. The static load-carrying capacity is mainly affected, but the fatigue strength is also somewhat reduced. Pitting corrosion and crevice corrosion *local* damage, which has a negligible impact on the static load-carrying capacity but can have devastating consequences for fatigue. The small areas of corrosion damage act as stress raisers, which can substantially lower the fatigue life. This is only applicable to unnotched steel members however. Several investigations have shown that as long as the rivet hole edges are unaffected, there can be substantial corrosion damage without causing a loss in the fatigue life of the riveted bridge members, cf. refs. [21] and [27]. Using a "weakest link" metaphor one can say that as long as the corrosion damage is not too severe, the rivet hole (read, the largest notch) will govern the fatigue life of a riveted bridge member.

Unlike a welded detail, a riveted connection is protected against corrosion at its weakest point (i.e. the rivet hole edge) by the rivet head. The clamping force keeps the joint tight and makes it impossible for water to seep in (at least in the vicinity of the rivet hole). The weakest point of a welded detail is the weld itself. The surface is totally exposed to the atmosphere, and any corrosion here will immediately lower the fatigue life.

4.6 Inspection and maintenance

A riveted railway bridge has to be inspected at regular intervals in order to operate safely. Special attention must be given to:

- Corrosion damages.
- Loose rivets.
- Faults and defects.
- Collision damages.
- Fatigue cracks.

If left unattended, each of these factors can significantly shorten the service life of a railway bridge. Normally, the railway authorities have inspection routines that detect these kinds of imperfections before the become critical, and if any fault is found, the right measure is immediately taken in order to maintain operation (e.g. repainting).

In Sweden, riveted railway bridges are thoroughly investigated every twelve years (in some cases every six years) to search for loose rivets (the so-called "rivet revision"). The main purpose of replacing loose rivets with new ones is to maintain the clamping force in the joints. There are many reasons for this:

- Better resistance against fatigue cracking.
- Smaller relative movements between interfacing plates (read, ensuring composite action).
- Less fretting.
- Smaller deformations.
- Less bearing pressure and less rivet shearing (i.e. the shearing forces will be transferred by friction rather than by bearing).
- A reduced risk of water penetrating between interfacing plates or under the rivet heads.

A special problem is how to detect fatigue cracks in good time before they have become "critical" (i.e. before the load-carrying capacity is reduced or before brittle fracture is initiated). There are several ways of determining the presence of fatigue cracks in a riveted railway bridge:

- Visual inspection.
- Ultrasonic testing.
- Acoustic emission.

Large fatigue cracks can be found by visual inspection. But how can we detect small fatigue cracks hidden under the rivet head? Many railway authority personnel argue for the use of acoustic emission as a non-destructive testing method in the field. However, there are some circumstances that speak for the use of ultrasonic testing:

- Acoustic emission is more suitable for welded structures where the fatigue cracks are already visible. The method can only confirm the *propagation* of a known crack.
- Acoustic emission is a method that requires a *traffic load* passing over the bridge.

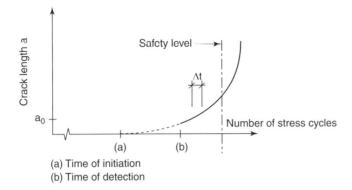

Fig. 4.13 The inspection interval must be chosen with reference to the crack propagation rate (i.e. the crack length as a function of the number of stress cycles) and a safety margin against failure.

- Ultrasonic testing can detect the presence of fatigue cracks *hidden* under rivet heads, whether or not these cracks are propagating (i.e. regardless of whether the structure is being subjected to traffic loading).
- Acoustic emission requires that the material and structural properties do not vary. Small *friction slips* can be expected to occur when a shear load is transferred in a riveted built-up section, and it is impossible to separate these sounds from those of the propagation of a crack.

Once a crack has been detected or has become visible, it normally requires a drastic change in the inspection routines. When the inspection personnel know the location of a "damaged" detail that should be kept under special surveillance, it is possible to concentrate efforts on this "critical" point. If the crack is allowed to propagate (which is not the normal procedure), the inspection interval must be adjusted to the crack propagation rate in order to ensure a "safe life", see Fig. 4.13.

Normally a fatigue crack is immediately stopped (temporarily) by drilling a hole at the tip of the crack. This has proved to be an effective procedure in many cases, although its use is primarily restricted to welded bridges. This is due to the fact that very few fatigue cracks have been found in riveted railway bridges over the years. If a fatigue crack is found, however, there could be less margin of safety (i.e. less time until complete fracture) if an ageing effect has affected the steel's material characteristics, cf. Fig. 4.13 and also the earlier discussion).

4.7 Reinforcement methods

There can be several reasons behind a decision to reinforce an existing riveted railway bridge, namely:

- Inferior load-carrying capacity.
- Corrosion damages.
- Large deflections (or excessive deformations).

Fig. 4.14 The configuration of a truss bridge before and after "arch reinforcement".

Fig. 4.15 A proposed method of reinforcing old riveted girder or truss bridges.

- Loose or faulty rivets.
- Fatigue cracks.
- Fear of brittle fracture.
- Collision damages.

The main means of performing an upgrading or retrofitting of a riveted railway bridge in Sweden over the years has been by adding extra plates to the overstressed regions or, in some cases, by replacing faulty rivets with high-strength friction-grip bolts. The use of welding as a connecting method when reinforcing a riveted railway bridge has more or less been "banned" due to the fear of fatigue and embrittlement.

In some cases, there have been examples of "concrete cladding", where a riveted steel bridge has been totally encased within a concrete "shell". Apart from increasing the static load-carrying capacity and decreasing the risk of fatigue cracking, this method has many advantages:

- The concrete protects the steel core against corrosion.
- The steel bridge members act as reinforcement for the concrete.
- There is an increased stiffness in the structure.

A more refined method is to strengthen old riveted bridges through "arch reinforcement". This method has been used successfully in the USA to rehabilitate old riveted truss bridges, cf. ref. [118]. The truss bridges have been reinforced by simply adding an arch to the superstructure, see Fig. 4.14.

The arch reinforcement method has proven to be a simple solution to the problem of reinforcing old riveted bridges without any major traffic disturbances.

Another method of reinforcing old riveted railway bridges, both girder and truss bridges, is to apply a steel trough on top of the longitudinal girders/stringers, see Fig. 4.15.

This method has a great potential, as it accommodates the urge of railway authorities to arrange ballast over the bridges as well as upgrading the existing riveted railway bridge structures, which in many cases deserve to be preserved for future use.

Chapter 5

Field Studies

5.1 General

A number of field investigations of riveted railway bridges were undertaken by the Department of Structural Engineering, Division of Steel and Timber Structures at Chalmers University of Technology in 1989–1992, cf. refs. [1], [3], [5] and [9]. In all, the field investigations resulted in studies of 15 riveted railway bridges, see Table 5.1.

These were the major concerns behind the decision to investigate the superstructure of these old riveted railway bridges:

- Their static load-carrying capacity is not known exactly. The main reason for this is that the quality properties of the steel are unknown, but also that (in some cases) there are doubts about the structural performance.
- What is the magnitude of the dynamic live load stresses?

Table 5.1 Riveted railway bridges investigated by the Department of Structural Engineering at Chalmers University of Technology during the four-year-period 1989–1992.

River	Nearest town/village	Structural type	Span length	Construction year
Lagan	Laholm	Truss/girder	37.2 + 15.8 m	1903–1909
Laholms Kvarnränna	Laholm	Girder	10.0 m	1909
Ångermanälven	Forsmo	Truss	2 × 58.5 + 104.0 + 42.0 m	1912
Gide Älv	Björna	Truss	2 × 27.1 + 34.9 m	1913
Hjuksån	Hällnäs	Girder	18.4 m	1922
Vindelälven	Åmsele	Truss	36.2 + 51.6 m	1923
Maltån	Åmsele	Girder	18.4 m	1923
Arvån	Åmsele	Girder	23.5 m	1923
Ume Älv East	Lycksele	Truss	72.0 + 41.4 m	1924
Ume Älv Channel	Lycksele	Girder	18.3 m	1924
Ume Älv West	Lycksele	Girder/truss	2 × 18.4 + 36.2 m	1924
Rusbäcken	Kattisavan	Girder	21.0 m	1926
Tåskbäcken	Söderfors	Girder	16.0	1927
Paubäcken	Åskilje	Girder	16.0 m	1927
Ume Älv	Åskilje	Truss	4 × 46.6 m	1928

74 Fatigue Life of Riveted Steel Bridges

Fig. 5.1 The riveted railway bridge over the Ångermanälven at Forsmo was investigated by the Department of Structural Engineering at Chalmers University of Technology in 1990 [1]. The bridge was built in 1912 and has a main (arch) span of 104 m.

- What effect does the inevitable corrosion damage have on the load-carrying capacity and remaining lifespan?
- Is there any other damage or defects that might have influence on the overall behaviour?
- Is the vertical deflection of each bridge during train passages tolerable?
- Is the lateral stiffness adequate?
- What is the dynamic amplification of the "static response"?
- The fracture toughness of the steel is unknown, and without that knowledge, how can we decide whether an observed defect is tolerable or not with reference to brittle fracture?
- Is the steel susceptible to ageing, i.e. is it a steel type which is known to have a large percentage of nitrogen and is therefore prone to becoming more brittle over time?
- What is the expected lifespan of each bridge according to a fatigue damage accumulation analysis?
- Are there any fatigue cracks present?

5.2 Visual inspection

The main and principal findings from the visual inspections of the riveted bridges, see Table 5.1, were:

- The nominal section dimensions were checked against the original design drawings. No differences were found.
- There was no major corrosion damage, i.e. pitting corrosion or crevice corrosion. A light surface corrosion could be noted, especially on horizontal or inclined surfaces.
- The bridge bearings were checked and were found to be functioning well despite the presence of debris and corrosion.
- A notable settlement in the ballast bed near the abutment was observed in some cases. The dynamic amplification of the live load stresses can be expected to be high (especially for the stringer and floor beams near the abutment) because of this settlement. This was also confirmed when performing strain measurements on the actual bridges.
- The track alignment was checked and it was found that the deviation from the ideal position never exceeded the tolerance value (100 mm) stipulated in the codes for the design of new railway bridges. In general the misalignment was less than 50 mm.
- No loose or faulty rivets were detected.
- No major damage or defect that reduces the fatigue life or increases the probability of brittle fracture could be found.
- No visible cracks in the riveted connections were observed.

5.3 Strain measurements

The main purpose of measuring live load strains by applying electrical resistance strain gauges to a bridge superstructure is to determine the general stress level, but equally important is the information that this provides on:

- Dynamic amplification.
- Stress range.
- Number of stress cycles.
- Strain rate.
- Secondary stresses.

The strain gauges should either be put in a region where one expects the highest stress level or in a region where the stress range is at its maximum. In the field investigations performed by the author, the method chosen was to only record a *discrete* analogous signal and not a continuous signal. The reasons behind this choice were:

- When performing a large series of strain recordings for many train passages over several days there would be difficulties in handling the output data if "continuous signal recording" was used.
- A low expectancy of a dynamic response, i.e. a contribution to the stress level from resonant vibration or forced vibration.

- Last, but not least, the available equipment made it preferable to choose a discrete signal recording.

The time interval between the strain recordings (read, the time increment) has to be chosen with care. In order to achieve good accuracy, and at the same time to maximize the amount of output data (within reasonable limits), the time interval Δt chosen was 0.5 s. This table illustrates the dependence of the accuracy of the strain readings, ε, on the time interval, Δt, for the measuring equipment:

Δt	ε	σ
0.1 s	$\pm 10 \cdot 10^{-6}$	± 2.1 MPa
0.5 s	$\pm 1 \cdot 10^{-6}$	± 0.21 MPa
1.0 s	$\pm 0.1 \cdot 10^{-6}$	± 0.021 MPa

The margin of error given by the strain recording equipment is thus ± 1 microstrains for a time interval of 0.5 s. Many other factors also influence the accuracy of strain recordings:

- Misalignment in the positioning of the strain gauges.
- Thermal deformations.
- Electrical interference from the overhead electrical distribution cables on the railway.
- Strain gauge cable length.
- Shielded or non-shielded strain cables.
- Systematic errors.
- Human errors.

Some typical results from the strain measurements taken during the field investigations are presented in Figs. 5.2–5.8 and Table 5.2. The theoretical curves are calculated by using a special computer program called "SUPERINF" [6], which was developed at the Department of Structural Engineering at Chalmers University of Technology. When using this computer program, the train set must of course be known with regard to axle spacing and axle loading in order to exactly determine this theoretical response. When there is a time difference between the calculated strain/stress response and the measured strain response, this is only due to the fact that the train's speed in the input data for the computer program differs from the actual value for the real train.

The main conclusions that can be drawn from the strain measurements performed during the field investigations (refs [1], [3] and [9]) are:

- The maximum stress levels for truss and girder bridges were generally about 42 MPa (200 microstrain) or less, even for the heaviest freight train loading.
- There was a surprisingly good correspondence between the theoretically calculated strain response and the actual measured response.
- In general, the measured strain values were lower than the predicted values.
- The dynamic amplification was in most cases found to be negligible.

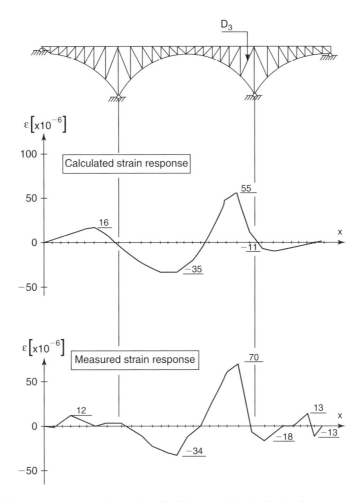

Fig. 5.2 The strain response in truss bar D_3 (theoretically calculated values compared with measured values) for a locomotive of 76 tons passing slowly step by step over the Forsmo Bridge [1].

– With reference to fatigue damage accumulation, the measured stress ranges proved to be much lower than the predicted values, and the number of stress cycles at the highest stress range were found to be much fewer than predicted.

5.4 Deflection

Load tests were carried out for the Forsmo and the Gide Älv Bridges in order to measure the deflections, see Tables 5.3 and 5.4. For the Forsmo Bridge it was possible to compare the structural performance with the initial behaviour back in 1912.

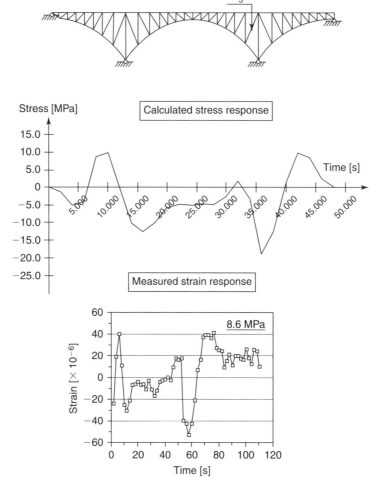

Fig. 5.3 The stress/strain response in truss bar D_3 (theoretically calculated values compared with measured values) for a heavy freight train (two locomotives, 76 tons each, and 21 wagons, 83–90 tons each) passing at 40 km/h over the Forsmo Bridge [1].

The main conclusions that can be drawn from the load tests are:

- In general, smaller deflections than the predicted values were measured.
- The design load in current practice (train load UIC 71) gives theoretical deflections that are significantly higher than those observed from both the static load test (in the 1912 load test) and the heaviest freight trains.
- There was fairly good agreement (though with both higher and lower values) between the static load test in 1912 and the equivalent load test in 1989.
- The measured deflections as well as the theoretically calculated values were significantly lower than the allowable values in the design code (i.e L/800).

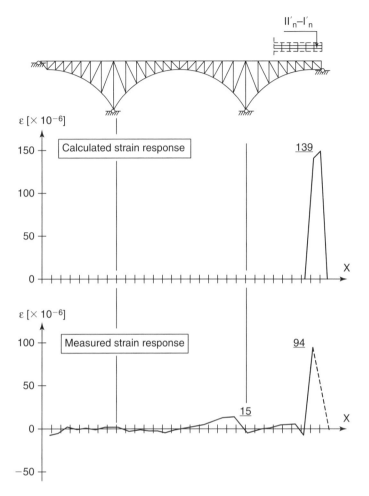

Fig. 5.4 The strain response in stringer $II'_n - I'_n$ (theoretically calculated values compared with measured values) for a locomotive of 76 tons passing slowly step by step over the Forsmo Bridge [1].

5.5 Dynamic amplification

In addition to the extensive strain measurements, the horizontal displacement at the roller bearings was measured at the Gide Älv Bridge. This was done in order to:

- Determine the magnitude of the vertical deflections (through the correlation between horizontal and vertical displacement).
- Confirm the low magnification of the static strain response through dynamic amplification that was found in the strain measurements.

Two displacement gauges were mounted at the roller bearings for the main span and for one of the side spans. The idea was to record (with a continuous signal this time) the

80 Fatigue Life of Riveted Steel Bridges

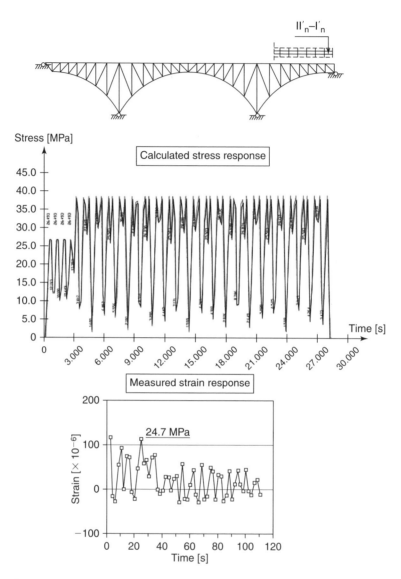

Fig. 5.5 The stress/strain response in stringer $II'_n - I'_n$ (theoretically calculated values compared with measured values) for a heavy freight train (two locomotives, 76 tons each, and 21 wagons, 83–90 tons each) passing at 40 km/h over the Forsmo Bridge [1].

horizontal displacement of the bearings as the trains were passing, see Fig. 5.9. There is a direct correlation between this horizontal movement and the vertical deflection, and, by using this procedure, the method of determining the vertical deflections was simplified.

Fig. 5.10 and Fig. 5.11 compare theoretically calculated horizontal displacements (dynamic forces, such as mass inertia forces or resonant vibration, are excluded) with

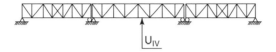

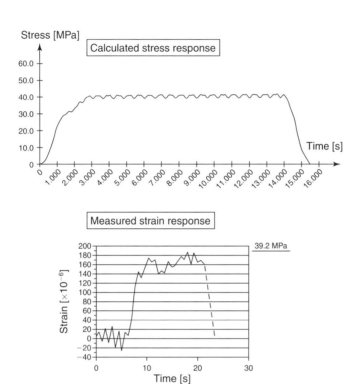

Fig. 5.6 The stress/strain response in truss bar U_{IV} (theoretically calculated values compared with measured values) for a heavy freight train (two locomotives, 76 tons each, and 21 wagons, 83–90 tons each) passing at 70 km/h over the Gide Älv Bridge [3].

the measured response for one of the side spans and for the main span of the Gide Älv Bridge.

These are the lowest calculated eigenfrequency of the two spans:

	Without traffic load	With traffic load
Main span	7.8 Hz	4.7 Hz
Side span	11.1 Hz	5.7 Hz

The main conclusions that can be drawn from the results of these horizontal displacement measurements are:

- The maximum measured displacement was approximately 10% below the calculated static value. In a simplified model, a structure is often assumed to be less rigid than the actual behaviour (factors such as friction at the roller bearings,

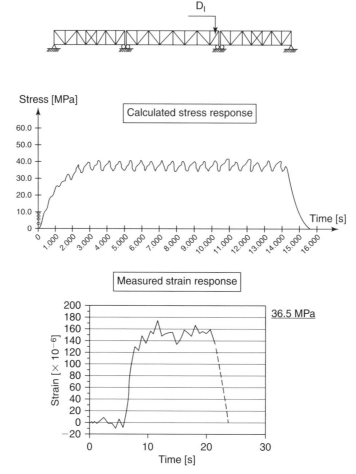

Fig. 5.7 The stress/strain response in truss bar D_I (theoretically calculated values compared with measured values) for a heavy freight train (two locomotives, 76 tons each, and 21 wagons, 83–90 tons each) passing at 70 km/h over the Gide Älv Bridge [3].

bending stiffness in joints and load distribution are excluded from the model), thus explaining why the "dynamic response" can be somewhat lower than the "static response".

- The "wave form" was more pronounced in the measured curve in comparison to the calculated static curve. This is due to the magnification through mass inertia forces when the train axles are passing the midspan, where the deflection (read, radius of curvature) is greatest.
- An "overlapping" vibration can be seen when comparing the measured curves with the theoretical ones. In particular it seems that this vibration is periodically recurring for the main span, cf. Fig. 5.11. The period of this vibration corresponds quite well to the eigenfrequency of the truss span ($f_1 = 4.7$ Hz). Despite the fact

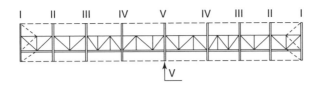
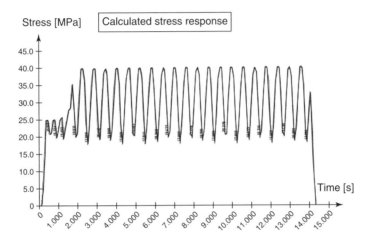
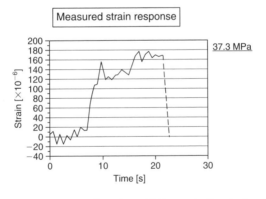

Fig. 5.8 The stress/strain response in floor-beam V (theoretically calculated values compared with measured values) for a heavy freight train (two locomotives, 76 tons each, and 21 wagons 83–90 tons each) passing at 70 km/h over the main span of the Gide Älv Bridge [3].

that these vibrations are emanating from resonance, they are, as can be seen, of smaller magnitude. It is more probably an effect from the uneven track surface rather than resonance between the train load and the truss bridge.

5.6 Crack control

Of the 15 riveted railway bridges that were investigated by the author (see Table 5.1), three bridges were checked more thoroughly for fatigue cracks than the others, namely

Table 5.2 Maximum strain values, expressed in microstrain, (measured values for a locomotive compared with theoretically calculated values) in three of the bridges investigated along the railway line between Hällnäs and Storuman [9].

	Horizontal tension bar at mid-span	Vertical floor-beam hanger	Stringer	Floor beam
Vindelälven (Åmsele)				
• A locomotive (76 tons) standing still	80	114	−61[1]	−102[1]
• A locomotive (76 tons) passing at 50 km/h	80	88	−60	−80
• Theoretically calculated static maximum strain	102	89	−[2]	−[2]
Ume Älv East (Lycksele)				
• A locomotive (76 tons) standing still	31	141	−162	−86
• A locomotive (76 tons) passing at 50 km/h	33	126	−108	−58
• Theoretically calculated static maximum strain	68	112	−206	−205
Ume Älv East (Åskilje)				
• A locomotive (76 tons) standing still	69	153	−121	−119
• A locomotive (76 tons) passing at 50 km/h	54	135	−86	−99
• Theoretically calculated static maximum strain	89	119	−196	−177

1) Negative values here because the strain gauges were put on the upper flange of the stringers and floor beams.
2) The design drawings of these girders were missing.

the Forsmo Bridge, the Gide Älv Bridge and the Laholm Bridge. In these bridges, ultrasonic testing was used to search for fatigue cracks at the rivet holes. In a riveted joint the fatigue crack normally originates from the edge of a rivet hole, and is initially hidden by the rivet head, thus making it impossible to detect the crack visually. The only method available to detect these kinds of cracks is ultrasonic testing. Every other crack detection method is more suited for welded construction. In ultrasonic testing a transmitter generates a sound vibration with a frequency of normally 0.5–10 MHz. This signal is either received on the other side of the investigated piece, or reflected back to the transmitter (an "echo-sounding" kind of method). This latter "impulse-echo method" is the most commonly used ultrasonic testing method. The time it takes for the "sound wave" to reflect back to the transmitter is directly proportional to the distance which has been travelled. A crack, or some other defect, makes the sound waves reflect earlier, resulting in a "default echo" on the screen in comparison to the ordinary signal.

The ultrasonic equipment used for the investigation of the bridges (Krautkrämer USK 7) had the capacity to detect cracks as small as 0.5 mm. To detect such small "reflection areas" requires perfect and homogeneous steel, which is definitively not the case for older steels. An inhomogeneous and, in some cases, loose layer of paint

Table 5.3 The load deflection of the Forsmo Bridge (theoretically calculated values compared with measured values) [1].

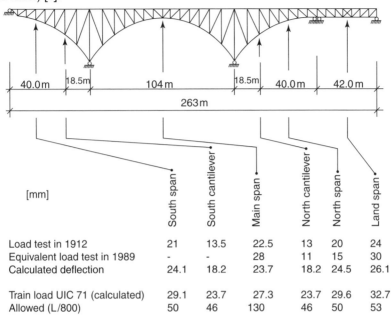

[mm]	South span	South cantilever	Main span	North cantilever	North span	Land span
Load test in 1912	21	13.5	22.5	13	20	24
Equivalent load test in 1989	-	-	28	11	15	30
Calculated deflection	24.1	18.2	23.7	18.2	24.5	26.1
Train load UIC 71 (calculated)	29.1	23.7	27.3	23.7	29.6	32.7
Allowed (L/800)	50	46	130	46	50	53

Table 5.4 The load deflection of the Gide Älv Bridge (theoretically values compared with measured values) [3]. The bridge has a main span of 34.9 m and two side spans of 27.1 m each.

[mm]	Side spans	Main span
A heavy freight train ("Stålpilen")	8.8	14.0
Calculated deflection	9.7	15.6
Train load UIC 71 (calculated)	12.3	19.6
Allowed (L/800)	34	44

makes detection of very small cracks even more difficult. Given these circumstances, the smallest crack length it was possible to detect in these three investigations was about 2 mm.

The riveted joints that were chosen for the crack investigations were:

- The midspan region and support region of stringers and floor beams.
- Truss members in tension at their connections.

An example of the rivets investigated (including the rivet holes) in a truss member connection is shown in Fig. 5.12.

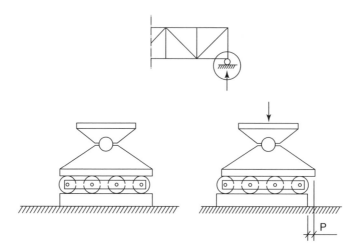

Fig. 5.9 During vertical loading of a truss bridge the superstructure is moved horizontally at the position of the roller bearings (in accordance with beam theory).

The number of rivets (and rivet holes) that were checked for fatigue cracking in the bridges were:

The Forsmo Bridge	268 rivets
The Gide Älv Bridge	260 "
The Laholm Bridge	48 "
	576 "

No indication of any cracks were detected in these bridges by the ultrasonic equipment, even though 576 rivets in total were tested.

5.7 Steel samples

Steel samples have to be taken from the bridges in order to determine, among other things, the steel's mechanical properties. The procedure is not that easy and many considerations have to be weighed before undertaking this task. There are, in principle, three different approaches that one could adopt:

- Take steel samples from areas/parts of members where the stresses are the lowest.
- Replace a whole bridge member (e.g. a secondary truss bar).
- Remove unstressed end parts of L-flanges.

This last method was successfully used on the riveted bridges investigated by the author. A triangular piece was cut from those places where the L-flanges had a free end, see Fig. 5.13.

The advantage of using this method is threefold:

- Many places in a bridge are available.

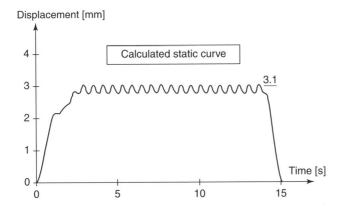

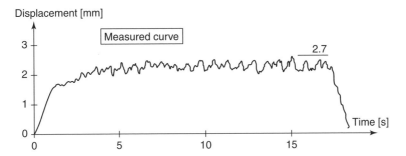

Fig. 5.10 The horizontal displacements of the roller bearing (theoretically calculated curve compared with measured curve) for a heavy freight train (two locomotives, 76 tons each, and 20 wagons, 83–90 tons each) passing at 80 km/h over the *side span* of the Gide Älv Bridge [3].

- The work of removing these parts is simple, not time-consuming and the railway traffic is not disturbed.
- The steel samples can be taken from locations evenly distributed over the bridge.

5.8 Material testing

Small test specimens were machined from the triangular steel samples that were taken from the bridges – see section 5.7 – in order to investigate:

- The steel's mechanical properties (i.e. yield point, tensile strength and elongation).
- Evidence of a possible ageing effect.
- Fracture toughness.
- Chemical content.
- Grain size.
- The distribution between ferrite and perlite.
- The presence of different slag inclusions.

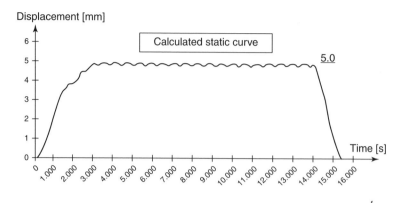

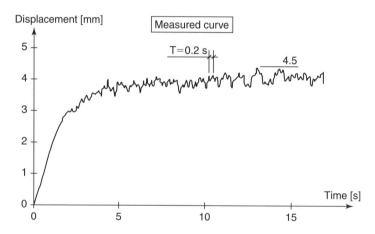

Fig. 5.11 The horizontal displacements of the roller bearing (theoretically calculated curve compared with measured curve) for a heavy freight train (two locomotives, 76 tons each, and 20 wagons, 83–90 tons each) passing at 80 km/h over the *main span* of the Gide Älv Bridge [3].

The results from these investigations are presented in the same order as the parameters and features listed above. The *tensile test* on the steel from the different bridges under investigation gave the following average values:

	Construction year	Yield point R_{eL} [MPa]	Tensile strength R_m [MPa]	Elongation A [%]
The Forsmo Bridge	1912	295	448	28.0 (A_7)
The Gide Älv Bridge	1913	282	396	28.5 (A_6)
Vindelälven (Åmsele)	1923	270	392	36.8 (A_5)
Ume Älv (Lycksele)	1924	304	439	33.7 (A_5)
Ume Älv (Åskilje)	1928	272	393	35.0 (A_5)

Fig. 5.12 The particular rivets chosen for crack control in a riveted truss joint [1]. Attention must be given to unequal load distribution between the rivets as well as secondary bending stresses.

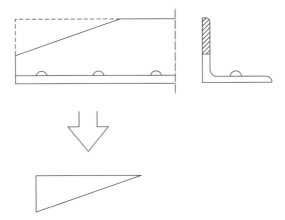

Fig. 5.13 Steel samples were taken from the free end parts of L-flanges.

During the tensile test, ageing behaviour was investigated using a method to create a fully developed effect in a reduced time. The tensile test on the test specimens was interrupted at a given elongation (5–10%). This was done in order to produce the cold-work deformation necessary to replicate a fully developed effect of ageing over time. The strained specimens were then put in a hot oven (+200°C) for one hour in order to accelerate the diffusion process. After the heat treatment, the tensile test

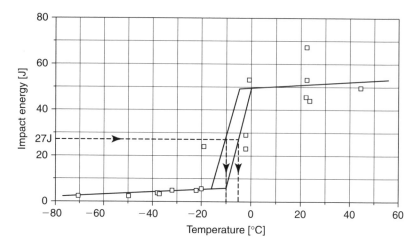

Fig. 5.14 Charpy V-notch impact test of steel specimens taken from the Forsmo Bridge [1].

was concluded. The tensile test performed in order to study the *ageing effect* gave the following results (average values):

	Yield point R_{eL} [MPa]	Tensile strength R_m [MPa]	Elongation A [%]
The Gide Älv Bridge	410	436	26.4
Vindelälven (Åmsele)	416	440	29.7
Ume Älv (Lycksele)	452	475	28.7
Ume Älv (Åskilje)	408	437	30.7

In comparison to the specimens from each bridge that were tested without any prior cold-work deformation and heat treatment, i.e. those in the original tensile test, the changes due to ageing were as follows:

	Yield point	Tensile strength	Elongation
The Gide Älv Bridge	+45%	+10%	−7%
Vindelälven (Åmsele)	+54%	+12%	−19%
Ume Älv (Lycksele)	+49%	+8%	−15%
Ume Älv (Åskilje)	+50%	+11%	−12 %

An impact notch test was performed on the steel from the Forsmo Bridge in order to determine its *fracture toughness* behaviour. Sixteen specimens ($5 \times 10\,\text{mm}^2$) were tested at temperatures ranging from +44 to −70°C, see Fig. 5.14 for the results.

The transition temperature (at 27 J) was found to be around −5 to −10°C. With reference to the lowest temperature likely to be experienced in the field – which is approximately −30°C – the fracture toughness, according to this standardized testing method, is far too low.

A *chemical analysis* of the steel gave these results for each bridges tested:

	Carbon [%]	Manganese [%]	Sulphur [%]	Phosphorus [%]	Nitrogen [%]
The Forsmo Bridge	0.15	0.34	0.021	0.061	–
Hjuksån	0.07	0.57	0.013	0.035	0.007
Vindelälven (Åmsele)	0.13	0.40	0.066	0.029	0.003
Maltån	0.17	0.44	0.055	0.028	0.004
Arvån	0.15	0.46	0.040	0.040	0.003
Ume Älv East (Lycksele)	0.12	0.45	0.046	0.030	0.004
Ume Älv Channel (Lycksele)	0.16	0.35	0.041	0.028	0.003
Ume Älv West (Lycksele)	0.11	0.41	0.043	0.022	0.003
Rusbäcken	0.15	0.47	0.035	0.049	0.002
Tåskbäcken	0.13	0.55	0.036	0.031	0.003
Paubäcken	0.10	0.43	0.025	0.024	0.003
Ume Älv (Åskilje)	0.11	0.45	0.039	0.038	0.003

A *metallurgical investigation* by microscope of the steel from the Forsmo Bridge showed a homogeneous grain distribution between ferrite and perlite, with a grain size of 30–60 μm, see Fig. 5.15. In one of the samples some large slag inclusions (>0.25 mm) were found.

5.9 Load-carrying capacity

5.9.1 General

Using the results of the field investigations and theoretical analysis, it was possible to classify the bridges in terms of their remaining fatigue life and their static load-carrying capacity. This classification is based on:

- A calculation of the design load with respect to the applicable structural design codes.
- The remaining fatigue life with reference to an estimate of the fatigue damage accumulation.
- A fracture mechanics analysis.

5.9.2 Design load

At the beginning of the twentieth century, when most of the riveted railway bridges that were still in use in the early 1990s were built, the design load (i.e. the highest allowable axle weight) was 18 tons at most. This original design load has gradually been adjusted over the years in order to permit higher allowable axle weights. This increase in load-carrying capacity has been mostly achieved by raising the allowable stresses in the design codes. But the dynamic amplification factor and the set-up of the theoretical train design load have also been altered.

The static load-carrying capacity for truss bars and girders in the Forsmo Bridge and the Gide Älv Bridge are presented in Figs. 5.16–5.19. The figures show the "utilization factor" with reference to the structural design code valid in the early 1990s. The

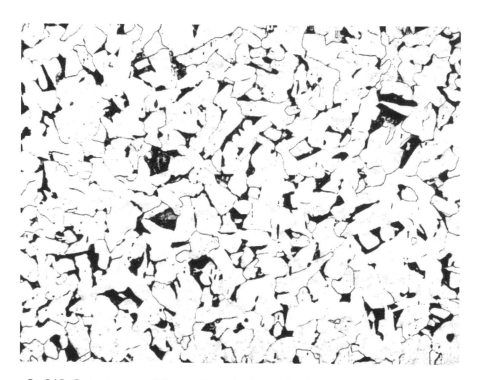

Fig. 5.15 Grain structure of the steel from the Forsmo Bridge. The lighter grains are ferrite and the darker ones are perlite. The share of perlite grains in this sample is about 5–10 %. The image is magnified 200 times [1].

Swedish railway code is harmonized with the railway code authorized by the international railway union (UIC) and the bridges have therefore been checked using train load UIC 71 (25 tons axle weight).

From the tension test the characteristic value of the yield point was derived, and this value served as the design stress. If the corresponding real design stress (read, steel quality) was not known, then it was impossible to determine the live load stress level that can be permitted.

5.9.3 Remaining fatigue life

The "dynamic" load-carrying capacity (read, fatigue life) was determined by estimating the fatigue damage accumulation in each bridge. By combining statistical data from the Swedish National Railways, see Fig. 5.20, with actual data of the particular railway line on which each bridge is situated, see Fig. 5.21, it was possible to determine, with a fairly high degree of accuracy, the loading history.

The question was then how to transform this "fatigue loading" into fatigue damage accumulation. There were still many uncertainties about how some parameters have changed during each bridge's operational life. Questions like these all contribute to a

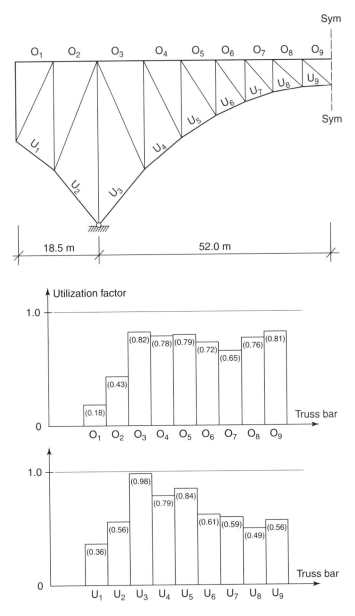

Fig. 5.16 Utilization factor for different truss bars in the main span of the Forsmo Bridge for a design load of UIC 71 (an axle weight of 25 tons) [1].

great uncertainty about how train loads in the past have been "transported" over the bridge, in particular with regard to:

- Axle weight.
- Axle spacing.

94 Fatigue Life of Riveted Steel Bridges

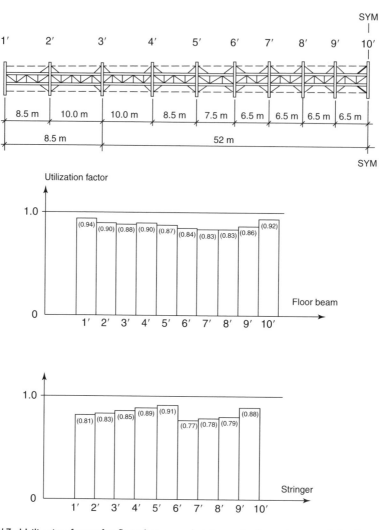

Fig. 5.17 Utilization factor for floor beams and stringers in the main span of the Forsmo Bridge for a design load of UIC 71 (an axle weight of 25 tons) [1].

– Bogie axles.
– Speed limits.

By using the statistical data given in Figs. 5.20–5.21 and transforming it into a number of "equivalent freight train passages", with the maximum axle weight allowed in the early 1990s, it is possible to arrive at an estimation of the fatigue damage

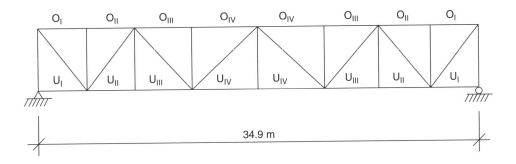

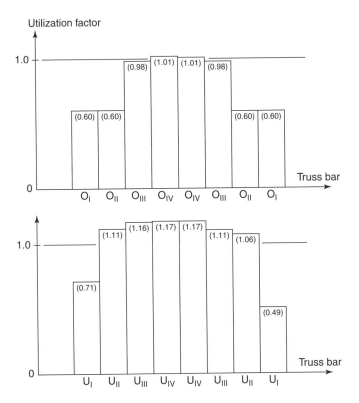

Fig. 5.18 Utilization factor for different truss bars in the main span of the Gide Älv Bridge for a design load of UIC 71 (an axle weight of 25 tons) [3].

accumulation that is on the safe side. This theoretical approach is based on the these facts:

– The maximum allowed axle weight in the early 1990s [at the time of the field investigations] is the highest permitted in the bridges' history.
– The fatigue life N (according to fatigue design curves) is inversely proportional to σ_r^m, where σ_r is the stress range (read, axle weight) and the exponent $m \geq 3$ (for $N < 5 \cdot 10^6$ $m = 3$ is used in Sweden).

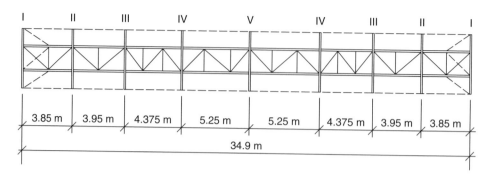

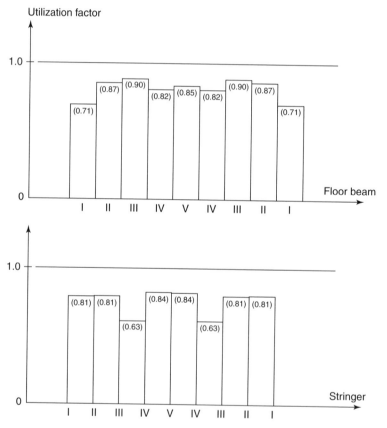

Fig. 5.19 Utilization factor for floor beams and stringers in the main span of the Gide Älv Bridge for a design load of UIC 71 (an axle weight of 25 tons) [3].

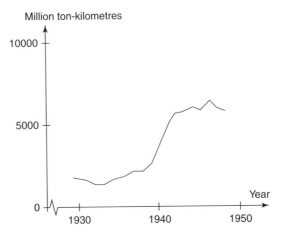

Fig. 5.20 Train loading of goods (in million ton-kilometres), with exception of "Lapland" iron ore, for the Swedish National Railways in the period 1929–1948 [1].

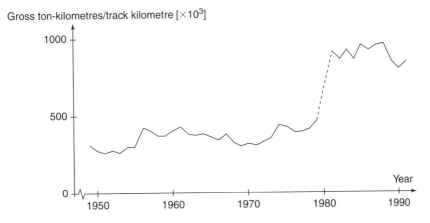

Fig. 5.21 Train loading (in gross ton-kilometres per track-kilometre) for the railway line between Hällnäs and Storuman in 1949–1991 (data from 1980 is missing) [9].

For a given train load, the fatigue damage accumulation D will be the highest for a minimum number of passages with a maximum allowed axle weight of each axle. This might perhaps need an explanation. Consider this simplified example:

$$\sigma_r = 100\ MPa$$

$$n = 2 \cdot 10^6 \quad \Rightarrow D = \frac{n}{N} \sim 2 \cdot 10^6 \cdot 100^3 = 2 \cdot 10^{12} \qquad (5.1)$$

$$N \sim \frac{1}{\sigma_r^3}$$

$$\sigma_r' = 80 \, MPa$$

$$n' = 2.5 \cdot 10^6 \left(\frac{100}{80} \cdot 2 \cdot 10^6\right) \Rightarrow D' = \frac{n'}{N'} \sim 2.5 \cdot 10^6 \cdot 80^3 = 1.28 \cdot 10^{12} \quad (5.2)$$

$$N' \sim \frac{1}{\sigma_r^3}$$

The change in the fatigue damage accumulation factor D can be expressed as a function of the quotient k between the stress ranges:

$$\sigma_r' = \frac{\sigma_r}{k}$$

$$n' = k \cdot n \Rightarrow D' = \frac{n'}{N'} \sim \frac{k \cdot n \cdot \sigma_r^3}{k^3} = \frac{D}{k^2} \quad (5.3)$$

$$N' \sim \frac{1}{\sigma_r'^3} = \frac{k^3}{\sigma_r^3}$$

As the stress range (read, axle weight) is reduced by the factor $1/k$, the number of passages increases by the same factor. But the fatigue life is then also increased, and by a factor of k^3 instead of just k. That is why the fatigue life accumulation factor D decreases as the stress range (read, axle weight) is decreased. The higher the weight for the "equivalent freight train passages", the higher and more conservative is the fatigue damage accumulation factor D. In the example given above, we see that as the stress range σ_r is decreased from 100 MPa to 80 MPa, the number of equivalent freight train passages is increased with a factor $k = 1.25$. The fatigue life N, however, increases by a factor of k^3, resulting in a decrease in the fatigue damage accumulation by a factor of $1/k^2$, equal to only 64% ($1/1.25^2$) of the original accumulation factor D.

The steps in calculating the fatigue damage accumulation factor D can be summarized as follows:

1. Using strain measurements or theoretical calculations of the strain/stress response for freight train passages, determine the stress range variation over time.
2. Determine the stress range for the heaviest freight train.
3. Using statistical data, determine the total train loading that has passed over the bridge during its life.
4. Transform this past train loading into a number of "equivalent freight train passages" by using the results from step 1 and 2.
5. Calculate the fatigue damage accumulation factor D equal to $\Sigma n_i / N_i$ for these "equivalent freight train passages".

If the factor D is found to be above 1.0, the results (according to the theory) tell you that the service life of the bridge is close to its end. But the bridge may still be in

Table 5.5 Estimated fatigue life after 1993 for the riveted bridges investigated by the author.

Bridge		Construction year	Remaining fatigue life (years)	Estimated annual increase in future traffic (%)
The Forsmo bridge		1912	31	5
The Gide Älv bridge		1913	>50	5
Hjuksån		1922	>50	5
Vindelälven (Åmsele)		1923	>50	5
Maltån		1923	>50	5
Arvån		1923	>50	5
Ume Älv East	(Lycksele)	1924	30	5
Ume. Älv Channel	(Lycksele)	1924	>50	5
Ume Älv West	(Lycksele)	1924	27	5
Rusbäcken		1926	>50	5
Tåskbäken		1927	>50	5
Paubäcken		1927	>50	5
Ume Älv (Åskilje)		1928	30	5

excellent condition without showing any signs of fatigue cracking. A thorough crack control investigation must confirm this fact before deciding about the future service life of the bridge.

If, however, the theoretical calculations of the fatigue damage accumulation factor D are below 1.0, the future "fatigue life" can be more "precisely" estimated. With respect to the traffic today and the expected increase in future traffic, an estimate can be made of the remaining theoretical fatigue life.

In Table 5.5 shows the results of theoretical calculations for the remaining fatigue life (i.e. until $D = 1.0$ is reached) of the riveted bridges investigated by the author.

Here one must add the caveat that the theory is simplified and based on assumptions, and cracks can very well be initiated before the fatigue damage accumulation factor equals 1.0. The effect of secondary bending stresses is always difficult to take into consideration.

5.9.4 Fracture mechanics analysis

In the following analysis, linear fracture mechanics theory is used to calculate the "fatigue life" of a floor beam. The particular beam chosen is an end beam (i.e. nearest to the abutment) at the Forsmo Bridge which, during the strain recordings, suffered the highest stresses (due to large dynamic amplification because of a very uneven ballast bed) for service loading. The largest non-detectable crack at a rivet hole that would not be picked up by ultrasonic testing is 2 mm, and that crack length is taken as an existing defect when calculating the "fatigue life", which is the number of stress cycles until this particular crack has propagated to a "critical value" (i.e. critical crack length) when brittle fracture is initiated.

The stress intensity factor K_I, at the tip of a fatigue crack from a rivet hole, is given by the following expression:

$$K_I = \sigma \cdot \sqrt{\pi \cdot a} \cdot f_6 \qquad (5.4)$$

K_I = stress intensity factor (mode I) [MPa$\sqrt{m}$]
σ = nominal stress [MPa]
a = crack length [m]
f_6 = a non-dimensional factor, which depends on rivet hole diameter and crack length

At a certain maximum value, K_I reaches a critical value (K_c) when unstable crack growth (i.e. brittle fracture) is taking place. K_c is called fracture toughness, and it is a material property that depends on temperature, strain rate, structural redundancy, defect position, environmental effects and other factors. From the expression given above for the stress intensity factor, equation 5.4, the critical crack length a_c can be derived:

$$a_c = \left(\frac{K_{Ic}}{\sigma \cdot f_6}\right)^2 \cdot \frac{1}{\pi} \tag{5.5}$$

The fracture toughness K_c can be derived using an empirical correlation between Charpy impact energy values (KV) and the fracture toughness level:

$$K_c = 11.76 \cdot KV^{0.66} \quad (KV < 45 \text{ J}) \tag{5.6}$$

At the lowest service temperature, the Charpy energy level was found to be in the neighbourhood of 4 J:

$$K_c = 11.76 \cdot 4^{0.66} = 29.4 \, MPa\sqrt{m} \tag{5.7}$$

The critical crack length can then be derived:

$$K_c = 29.4 \, MPa\sqrt{m}$$

$$\sigma = 72.5 \, MPa \quad \Rightarrow a_c = \left(\frac{29.4}{72.5 \cdot 1.1}\right)^2 \cdot \frac{1}{\pi} = 43 \, mm \tag{5.8}$$

$$f_6 = 1.1 (a/r \approx 4)$$

Some conditions must be fulfilled for the validity of linear fracture mechanics:

$$a, t > 2.5 \cdot \left(\frac{K_I}{\sigma_y}\right)^2 \tag{5.9}$$

a = crack length
t = plate thickness
K_I = stress intensity factor
σ_y = yield stress

In fatigue loading, the plastic zone at the crack tip is smaller than for static loading, and therefore the expression given above can be approximately set to:

$$a, t > 1.0 \cdot \left(\frac{K_I}{\sigma_y}\right)^2 \tag{5.10}$$

With the actual values:

$$a = 43\,mm, t = 10\,mm > 1.0 \cdot \left(\frac{29.4}{295}\right)^2 = 9.9\,mm\ OK! \qquad (5.11)$$

By using Paris' law it is possible to calculate the number of cycles (read, lifespan) needed for the crack to propagate from the non-detectable size of 2 mm to the critical crack length. In the following calculations this critical crack length is assumed to be 25 mm taking into account that the rivet hole in the lower flange of the floor beam is located near the plate edge. Furthermore, one can expect a large scatter in material properties "ancient" steel. The stress intensity variation parameter ΔK is calculated at the "beginning" (at a crack length of $a_0 = 2$ mm) and at the "end" (at a critical crack length of $a_c = 25$ mm) of the crack propagation process before brittle fracture:

$$\frac{da}{dN} = C \cdot \Delta K^m \qquad (5.12)$$

$\frac{da}{dN}$ = crack propagation rate [m/cycle]
C = constant
ΔK = stress intensity variation [MPa$\sqrt{m}$]
m = constant

$$\Delta K_0 = \Delta\sigma \cdot \sqrt{\pi \cdot a_0} \cdot f_6 = 72.5 \cdot \sqrt{\pi \cdot 2 \cdot 10^{-3}} \cdot 2.5 = 14.4\,MPa\sqrt{m} \qquad (5.13)$$

$$\Delta K_c = \Delta\sigma \cdot \sqrt{\pi \cdot a_c} \cdot f_6 = 72.5 \cdot \sqrt{\pi \cdot 25 \cdot 10^{-3}} \cdot 1.2 = 24.4\,MPa\sqrt{m} \qquad (5.14)$$

Typical approximate values of the constants C and m are used in order to calculate the crack propagation rate for the two different circumstances:

$$\left(\frac{da}{dN}\right)_0 = C \cdot \Delta K_0^m = 1.832 \cdot 10^{-13} \cdot 14.4^3 = 5.5 \cdot 10^{-10}\,m/cycle \qquad (5.15)$$

$$\left(\frac{da}{dN}\right)_c = C \cdot \Delta K_c^m = 1.832 \cdot 10^{-13} \cdot 24.4^3 = 26.6 \cdot 10^{-10}\,m/cycle \qquad (5.16)$$

In order to calculate exactly how many cycles are needed for the crack to propagate from 2 mm (a_0) to 25 mm (a_c), the expression for the crack propagation rate must be integrated. But first a polynomial expression must be adjusted to fit the change in $\sqrt{a} \cdot f_6$. As a simplified solution, one can assume the crack propagation rate to be constant and to take its highest value (at $a_c = 25$ mm):

$$\left(\frac{da}{dN}\right)_c \cdot N = \Delta a \qquad (5.17)$$

$$N = \frac{\Delta a}{(da/dN)_c} = \frac{(25-2) \cdot 10^{-3}}{26.6 \cdot 10^{-10}} = 8.6 \cdot 10^6\ cycles \qquad (5.18)$$

102 Fatigue Life of Riveted Steel Bridges

Given a safety factor of 4, which is chosen with reference to inspection routines and scatter in the actual data, the number of cycles is:

$$N_d = \frac{8.6 \cdot 10^6}{4} = 2.15 \cdot 10^6 \; cycles \qquad (5.19)$$

In fatigue damage accumulation calculations, the sum $D=1$ was found to be at $8 \cdot 10^5$ cycles. The conclusion of this fracture mechanics analysis is therefore that you would expect the final fracture to be a ductile failure (i.e. when the static tension load-carrying capacity is insufficient) rather than a brittle failure.

Chapter 6
Full-scale Fatigue Tests of Riveted Girders

6.1 General

When the old riveted railway bridge over the Vindelälven at Vännäsby was replaced by the railway authorities in 1993 it was decided to take the opportunity to carry out an extensive full-scale fatigue testing programme of the stringers and floor beams of the bridge, a task that was given to the Department of Structural Engineering at Chalmers University of Technology. This bridge was one of the oldest riveted railway bridges that had been in service during the twentieth century, and at the beginning of the 1990s it was decided that it had reached the end of its technical life. By performing a full-scale fatigue test on the stringers from the bridge, it was possible to achieve several results that could help to answer questions about the true lifespan of this particular bridge and other similar bridges. Questions that would be answered include:

- What is the remaining fatigue life?
- What is the fatigue damage accumulation?
- What is the fatigue fracture behaviour?
- Where are fractures located?
- What are the implications for the structural design codes, which are generally are based on fatigue tests performed on small riveted details?
- Is there an agreement between actual fatigue lives and the predicted lifespan according to a fatigue damage accumulation analysis?
- Which are the fatigue critical joints and connections?

6.2 Test specimens

In total 36 stringers and nine floor beams were removed from the old riveted railway bridge over the river Vindelälven outside the village of Vännäsby, located approximately 20 kilometres west of the city of Umeå. This railway bridge, built in 1896, was replaced in August 1993 because it had insufficient load-carrying capacity and, over the years, there were numerous occurrences of loose and fractured rivets. In addition, there had also been some complaints about "swaying tendencies" when trains were passing over the bridge. After consulting the design drawings and after verification of the actual bridge construction in the field, the author found that some important sway-bracing members and braking stiffeners were missing. They had not been taken away, but are missing as a result of inadequate knowledge about proper construction

104 Fatigue Life of Riveted Steel Bridges

Fig. 6.1 The old riveted railway bridge over the Vindelälven at Vännäsby was replaced in August 1993 by a continuous trough/girder composite construction.

detailing at the time when the bridge was designed. In regard to the "faulty" rivets (cf. Fig. 4.5 in section 4.2), when the bridge was built at then end of the nineteenth century it was not common practice to use pneumatic riveting machines. The preferred riveting method in the field at that time was to use hand hammers, and this can explain, to some extent, why the rivets are "inferior" when in comparison to those on riveted bridges of a later date. The swaying tendencies of the bridge could also very well be a contributing factor, adding secondary stresses and bending of the bridge members. The bridge had many years of service loading, and this of course is another important parameter that explains why the riveted connections were slowly "deteriorating".

The bridge had three simply supported "arch type" truss spans, each 71.2 metres in length, see Figs. 6.2–6.3.

Each truss span has 13 floor beams supporting 24 (2 × 12) stringers. The stringers are 5.933 metres in length and placed at a distance of 2.0 metres apart, see Fig. 6.4.

In December 1993 the girders chosen for the full-scale fatigue test programme were separated from the bridge before the rest of the bridge was demolished. Twenty-four stringers were separated one by one, and the rest, 12 stringers and nine floor beams, were detached as three separate "packages", each comprising three floor beams and four stringers, see Fig. 6.5.

The stringers have a built-up I-section consisting of an 8 mm thick web plate with four L-flanges (L 115 × 77 × 11), all riveted together with 20 mm rivets with a spacing of $c = 120$ mm in the longitudinal direction, see Fig. 6.8.

Spaced throughout the length of the detached stringers, see Fig. 6.9, there are four vertical stiffeners, each consisting of two vertical L-profiles (L 75 × 62 × 8), one on each side of the web plate.

Full-scale Fatigue Tests of Riveted Girders 105

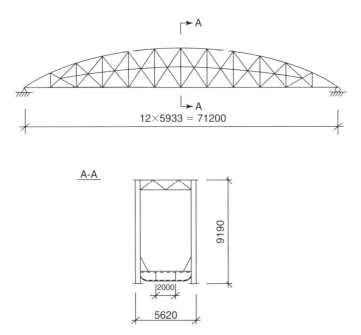

Fig. 6.2 Elevation of one of the three simply supported arch truss spans and a cross-section at the midspan of the old riveted railway bridge over the Vindelälven at Vännäsby. When this bridge was in service it was the eighth largest railway bridge in Sweden.

Fig. 6.3 The old railway bridge over the Vindelälven at Vännäsby (photo taken in June 1989).

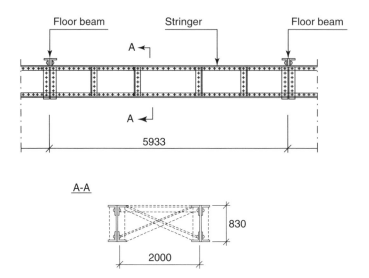

Fig. 6.4 Elevation and cross-section of the stringers.

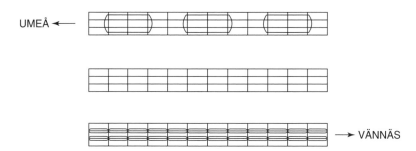

Fig. 6.5 Plan of the stringer/floor beam system of the railway bridge over the Vindelälven at Vännäsby. The stringers and floor beams chosen for the fatigue test programme are encircled in the plan showing the three spans.

Attached to the lower flange, at the position of the inner vertical stiffeners, are the "remains" of wind-framing cross-diagonals, cf. Figs. 6.6 and 6.7. The only reason why these diagonals were attached to the stringers in the original structure was that they had low bending stiffness – without this attachment the cross-diagonals would deflect too much. The cross-diagonals are either a single L-profile ("Type I") or a double L-profile ("Type II"), depending on their position in the horizontal wind-framing truss, see Figs. 6.10–6.12.

Nine stringers were chosen for a first series of fatigue tests, see Fig. 6.13. The rest of the stringers and floor beams were put aside to be tested at a later stage.

The nine stringers to be tested were carefully examined before the start of the fatigue test. The result of this visual examination was as follows:

Fig. 6.6 A photo taken of the work in December 1993 when the stringers in the span nearest Vännäsby (in the direction of Vännäs) were detached by flame cutting.

Fig. 6.7 Some of the detached stringers awaiting transport to Chalmers University of Technology in Göteborg.

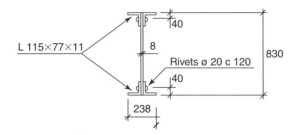

Fig. 6.8 Dimensions of the built-up I-shaped stringers.

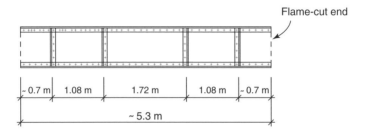

Fig. 6.9 Elevation of the detached stringers showing the position of the vertical stiffeners.

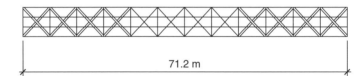

Fig. 6.10 Plan of the wind-framing system.

- No severe corrosion damage was found (i.e. pitting corrosion or major surface corrosion), although extensive areas of the web plate surface were corroded. In particular, the stringers on the "B-line" (cf. Fig. 6.13) had extensive surface corrosion on the web plate facing "out" into the open (probably a result of the prevailing wind direction over the years). The flanges were not as affected as the web plate – at most, the second layer of cover paint was missing in small areas on top of the compression and tension flanges.
- No fatigue cracks could be found.
- No loose rivets could be detected. The rivets were checked by lightly tapping a small hammer on the side of the rivet head.
- No other faults or defects could be found.

6.3 Test set-up and testing procedure

The stringers were tested by four-point bending in order to simulate a passage by a bogie-axle pair. The distance between the loading points was set to 2.0 metres, while the

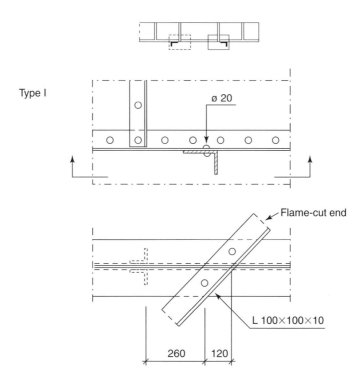

Fig. 6.11 Position of cross-diagonal Type I at the lower flange of the stringers.

Table 6.1 Stress range levels chosen for the nine stringers tested.

Stringer	Support distance [m]	P_{min} [kN]	$\sigma_{min}^{1)}$ [MPa]	P_{max} [kN]	$\sigma_{max}^{1)}$ [MPa]	P_1 [kN]	$\sigma_r^{1)}$ [MPa]	$R = \dfrac{\sigma_{min}}{\sigma_{max}}$
6B	5.0/4.0	29.4/44.1	12.8	121.6/182.4	52.8	92.2/138.3	40	0.24
7A	5.0/4.0	29.4/44.1	12.8	121.6/182.4	52.8	92.2/138.3	40	0.24
12A	4.0	44.1	12.8	251.4	72.8	207.3	60	0.18
2A	5.0	29.4	12.8	213.6	92.8	184.2	80	0.14
3B	5.0	29.4	12.8	213.6	92.8	184.2	80	0.14
5B	5.0	29.4	12.8	213.6	92.8	184.2	80	0.14
4B	5.0	88.3	38.4	318.7	138.4	230.4	100	0.28
5A	5.0	88.3	38.4	318.7	138.4	230.4	100	0.28
10B	5.0	88.3	38.4	318.7	138.4	230.4	100	0.28

1) Based on net section area and excluding the dead weight stress. The stresses are derived at the outermost edge of the beam (i.e. $\sigma_r = M_r/W_{net}$) instead of at the actual rivet locations.

distance between the supports was either 5.0 or 4.0 metres, see Fig. 6.14. The reason behind the decision to alter the span length from 5.0 to 4.0 metres (for three stringers) was to increase the flexural rigidity, and therefore reduce the vertical deflection. The capacity of the pulsating jacks increases as the stroke length decreases.

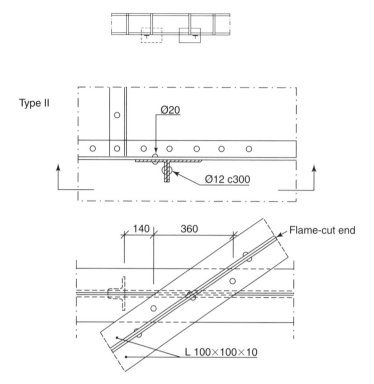

Fig. 6.12 Position of the cross-diagonal Type II at the lower flange of the stringers.

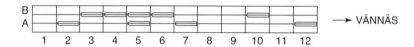

Fig. 6.13 These nine stringers were chosen for the first series of fatigue tests from the 24 stringers in total that were dismantled from the end span nearest Vännäs (cf. Fig. 6.5).

Two 60-ton hydraulic jacks were used to generate the pulsating load. The pulsating machine was of German origin (Losenhausenwerk) and had an oil volume of 300 cm^3 when pulsating. Their loading range capacity was between 5 and 65% of the maximum static loading capacity of the hydraulic jacks, i.e. a minimum load of 3 tons (29.4 kN) and a maximum load of 39 tons (382.6 kN). The stress range levels (i.e. the maximum nominal bending stress variation) chosen for the fatigue test were 40, 60, 80 and 100 MPa, and the stress ratio R varied between 0.14 and 0.28, see Table 6.1.

The stringers were stabilized with horizontal struts (bracings) at three points, at the two loading points on the top flange and at the midspan on the lower flange, see Fig. 6.15.

The upper two bracings were placed in order to ensure lateral stability during vertical loading, while the lower one should hamper the expected lateral secondary bending that takes place after fatigue crack initiation in one of the lower tension flanges.

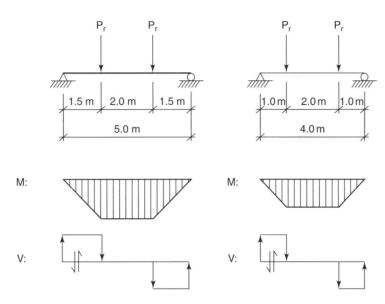

Fig. 6.14 Loading positions and the two choices of span lengths that were used in the full-scale fatigue test series.

It was intended to load stringer 12A (see Table 6.1) simultaneously with a second stringer (at the stress range of 60 MPa), as had been the case for the first two stringers tested, namely 6B and 7A. In order to meet the expected increase in vertical deflection (read, stroke length of the hydraulic jacks) the span length was decreased. This step proved to be inadequate though, due to the increased deformations in the rig itself. Consequently, the fatigue testing at $\sigma_r = 60$ MPa had to be performed one stringer at the time.

There are three critical locations in the bottom flange where one can expect possible fatigue cracking:

- In the constant-moment region between the loading points. The rivets in the "neck" here are not transferring any shear load, it is the rivet hole itself that is the stress raiser.
- Outside the constant-moment region, i.e. in the end parts of the stringers. The rivets in the "neck" here are transferring a shear load in this constant shear force region. The bending moment in this region varies, however, between zero and M_{max}, and therefore the rivets closest to the loading point are expected to be critical.
- At the flange-rivet connections for the cross-diagonals. These rivet locations vary, depending on whether the cross-diagonals consist of one or two L-profiles, cf. Figs. 6.11 and 6.12. A certain composite action is to be expected by "Type I" (a single L-profile), since these particular rivets are transferring a shear load (of small magnitude though). The rivets in cross-diagonal "Type II" (two L-profiles

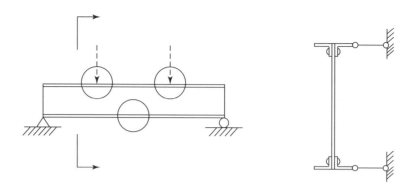

Fig. 6.15 Stabilizing points during pulsating loading.

Fig. 6.16 The loading rig in the laboratory. The rig was designed to carry four hydraulic jacks working at the same time, i.e. two stringers could be loaded simultaneously.

joined together) are, together with the L-profiles, acting somewhat differently and the shear load to be transferred is probably even smaller for these rivets. In both cases though, the rivets are located at or near the constant-moment region, see Fig. 6.17.

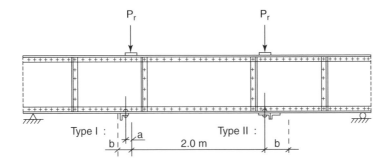

Fig. 6.17 The location (in mm) of the flange-rivet connections for the cross-diagonals:

	Type I	Type II
a	120	0
b	240	360

Table 6.2 The results from the full-scale fatigue testing.

Stress range $\sigma_r^{1)}$ (MPa)	Stress ratio R	Stringer number	Number of cycles to crack detection	Additional cycles	Failure
40	0.4	6B	–	–	(20 000 000)[2]
40	0.24	7A	–	–	(20 000 000)[2]
60	0.18	12A	–	–	(10 000 000)[2]
80	0.14	2A	5 893 350	102 520	5 995 870
80	0.14	3B	2 375 500	0	2 375 500
80	0.14	5B	6 370 440	114 900	6 485 340
100	0.28	4B	1 488 920	148 950	1 637 870
100	0.28	5A	2 038 660	146 240	2 184 900
100	0.28	10B	2 027 250	0	2 027 250

1) Based on net section area, the stresses are derived at the outermost edge of the beam (i.e. $\sigma_r = M_r/W_{net}$) instead of the actual critical rivet locations.
2) The test was discontinued after a given number of load cycles.

The available frequency capacity of the pulsator depends on the loading range (read, stroke length of the hydraulic jacks). Even for a small loading range, there is an upper limit of 360 cycles/min (i.e. 6 Hz).

6.4 Test results

6.4.1 Fatigue loading

The results of the full-scale fatigue testing of nine stringers taken from the railway bridge over the Vindelälven at Vännäsby are presented in Table 6.2 (cf. also Table 6.1) and Fig. 6.18.

The cracks have either started at the flange-rivet connections (Type I or II according to Figs. 6.11, 6.12 and 6.17) or at a "neck" rivet in the constant-moment region near the midspan, see Table 6.3.

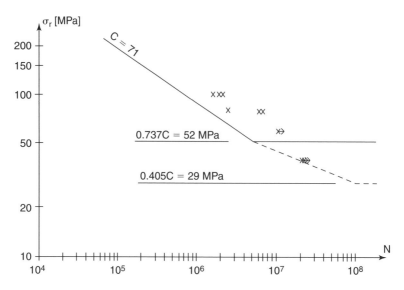

Fig. 6.18 The results given in a Wöhler diagram (number of cycles until failure as a function of the stress range). The fatigue design curve for riveted connections (C=71) in the Swedish National Steel Design Code (equal to category D according to AASHTO and equal to ECCS category detail 71) is shown for comparison.

Table 6.3 The location of crack initiation.

Stringer number	Stress range σ_r (MPa)	$\sigma_r^{1)}$ (MPa)	Crack initiation Flange rivet connection type I/type II	Neck rivet
2A	80	80	Type II	–
B3	80	80	II	–
5B	80	72,3	–	340 mm from mid-span
4B	100	100	II	–
5A	100	92	I	–
10B	100	100	II	–

1) The actual stress range at the rivet location of crack initiation.

The crack propagation scenario is the same in each case (i.e. irrespective of whether the crack starts at a flange rivet connection or at a "neck" rivet in the constant-moment region):

1. First, a fatigue crack is initiated at a rivet hole location in one of the two L- profiles at the lower tension flange.
2. When the fatigue crack has severed this L-profile, there is a temporary arrest until a second crack is formed in the second "tension string" (read, second L-profile).
3. In the meantime – during the arrest of the first fatigue crack – a crack has been initiated in the web plate, and when this crack has propagated to a certain length

above the vertical legs of the flange L-profiles, a crack is formed in the second L-profile.
4. When this L-profile is also broken, the dynamic load-carrying capacity is close to its end. Just a few more cycles are required to propagate the crack in the web a considerable distance upwards (25–30 cm or so).

The point of origin for the first initial crack in the first fractured L-profile has differed though. In the case of a crack initiation in a flange-rivet connection, the crack was initiated at the rivet hole in the outstanding leg of the L-profile, while in the case of the crack initiation in the constant-moment region (stringer 5B), the crack was initiated at one of the neck-rivet holes, i.e. 40 mm up into the web plate. In the former case, i.e. a crack initiation in a flange-rivet connection, the crack "jumps" one step towards the midspan (to the nearest neck-rivet) when passing from the first L-profile to the second, see Figs. 6.19–21.

In the case of a crack initiation at a neck-rivet hole located in the constant-moment region (stringer 5B), the crack formed in one L-profile and propagated to the other L-profile and the web plate at the *same* location. The reason for the "jump" in the former case (fatigue crack propagation at a flange-rivet connection) is that the region is overstressed due to the presence of the crack in the first L-flange, and the closest defect (read, rivet hole) towards the midspan (i.e. towards the maximum-moment region) is the neck-rivet next to the initial flange crack. Furthermore, "tension string 1" (i.e. the first L-profile) has lost its "anchorage" at the location of the flange-rivet and this creates large shear forces to be transferred by the neck-rivet next to the initial crack. Additional secondary lateral bending forces (due to the eccentricity of the normal forces in the lower flange) also increase the stresses in this region.

As has been noted earlier in this section, the crack propagation rate in the different parts of the lower flange (i.e. the two L-profiles and the web plate) differed markedly. The propagation rate was slow and steady for the first L-profile (i.e. where the first crack was initiated) and for the web plate (*before* crack initiation in the second profile), but fast and accelerating for the second L-profile as well as for the web plate once total failure was near, see Figs. 6.22–23.

As can be seen in Figs. 6.22–23 a considerable number of loading cycles are required after the first L-profile fractures completely until the second L-profile also fractures (cf. also Table 6.2). When monitoring the crack propagation in the different member parts of the bottom flange, it was found that there was an arrest in crack propagation when the first L-profile had completely fractured. A substantial number of loading cycles was required before initiation of a fatigue crack in the second L-profile was detectable. Through the composite action of the different member parts joined together, the stringers show a considerable inherent structural redundancy, which makes the fracture behaviour ductile (before the final failure) and the crack propagation extended in time, see Table 6.4.

In five of six fractured stringers (see Table 6.3) the initiation of the first fatigue crack was at a flange-rivet cross-diagonal connection. The expected crack location, taking the loading position into consideration (see e.g. Fig. 6.14), was somewhere in the lower flange between the loading points (i.e. at a neck rivet). Only in one case, however, was the fatigue crack initiated in this midspan region (stringer 5B). What then is the reason behind the fact that in most of these tests a secondary connection, located

116　Fatigue Life of Riveted Steel Bridges

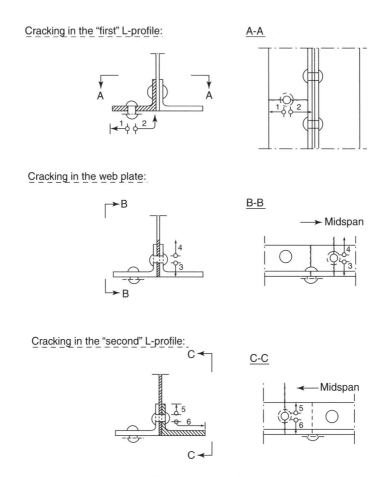

Fig. 6.19 Fatigue crack propagation after initiation at a flange-rivet location (Type I or II), i.e. for all stringers except number 5B. The crack initiation was always at the flange-rivet nearest to the midspan.

in the outer parts of the stringer, is governing the fatigue life and not the neck-rivets in the midspan region? The answer can not be the absence of transferring shear forces in the midspan region, because the neck-rivets nearest to the loading point in the outer parts transfer a shearing force equal to the applied jack load at the same time as they are experiencing almost the same bending moment as in the midspan region. The shear force in the flange-rivet connection is negligible in comparison, so there must be other explanations, for example:

– A flange-rivet connection Type II has a rivet located exactly 1.0 metre from the midspan, i.e. directly under the loading point. The stress range here is thus exactly the same as the maximum stress range for the beam. The neck-rivets are located 40 mm up in the web and therefore experience a slightly lower stress level

Fig. 6.20 The fatigue crack at failure for the first L-profile (i.e. where the crack was initiated at the flange-rivet connection) in stringer 5A.

than the outermost fibre of the beam, cf. for example stringers 3B and 5B in Table 6.3.
- In the flange-rivet connections the rivets are most probably hand riveted (we can assume this with a greater certainty than for the other rivets in the structure). Apart from giving an expected inferior performance, the resulting clamping force can also be expected to be lower when this technique is used.
- The rivet holes in these flange-rivet connections are also punched rather than drilled.

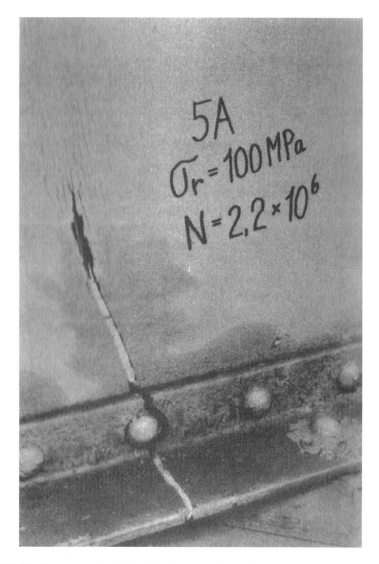

Fig. 6.21 The fatigue crack at failure for the second L-profile in stringer 5A.

- One can also expect that checks on the outcome of the riveting process for the flange-rivet connections may have been less thorough, due to the fact that the connections are only secondary (their only purpose, as has been stated before, is to reduce the vertical deflection of the cross-diagonals).

The final fatigue fracture (read, failure) for both cases, when the fatigue crack was initiated at a flange-rivet connection or at a neck-rivet in the midspan region, has been defined as the complete fracture of the lower tension flange (i.e. both L-profiles

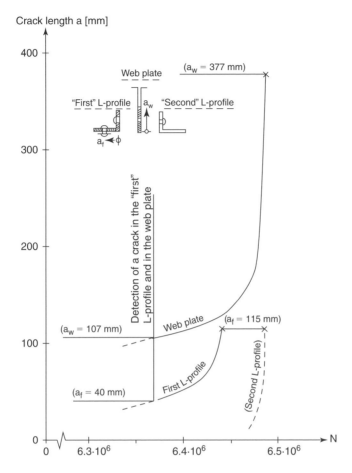

Fig. 6.22 Crack propagation in the first L-profile and in the web plate for stringer 5B (crack initiation at a neck-rivet).

completely fractured). The stringers have all been able to carry the prescribed cyclic load until the final fracture of the second L-profile. The vertical deflection as well as the secondary lateral bending (of the partially cracked tension flange) has, however, been excessive in these latter stages.

The results from the full-scale fatigue testing can also be used to study the effect of:

- Stress ratio R (i.e. $\sigma_{min}/\sigma_{max}$).
- Location in the original bridge (longitudinal position 1–12 and transverse position A or B, cf. Fig. 6.13).

120 Fatigue Life of Riveted Steel Bridges

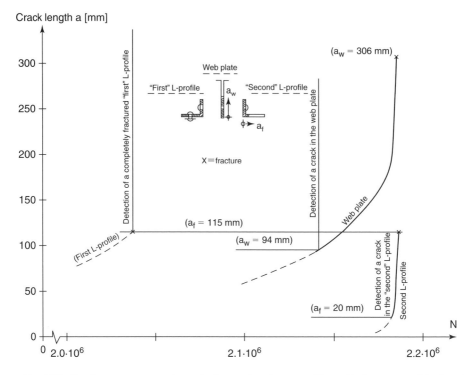

Fig. 6.23 Crack propagation in the web plate and in the second L-profile for stringer 5A after the first L-profile had fractured completely, cf. also Figs. 6.20–21.

Table 6.4 Structural redundancy of the stringers illustrated by the number of loading cycles in between different partial fractures being detected.

Stringer number	Stress range σ_r (MPa)	Number of cycles until:		ΔN
		Complete fracture in the "first" L-Profile	Detection of a crack in the "second" L-Profile	
2A	80	5 893 350	5 995 870[1)]	102 520
3B	80	–	2 375 500[1)]	–
5B	80	6 441 250	6 485 340[1)]	44 090
4B	100	1 539 560	1 637 100	97 540
5A	100	2 038 660	2 181 600	142 940
10B	100	–	2 027 250[1)]	–

1) A complete fracture! Due to the very fast crack propagation rate observed in the "second" L-profile (only a couple of thousand cycles or so from initiation to a complete fracture for stringers 4B and 5A), ΔN in this table is close to the actual number of cycles between a complete fracture in the "first" L-profile and an *initiation* of a crack in the "second" L-profile.

The influence of these "parameters" can be summarized as follows (taking into consideration though that this is only based on six fractured stringers):

Parameter	Influence (comment)
Stress ratio R	A slight trend for the two largest stress ranges can be observed. The reduction in fatigue life for $\sigma_r = 100$ MPa is too strong just to be explained by the "law" defined by the slope of the fatigue design curve that is normally adopted (i.e. $m = 3$). The stress ratio is 0.28 for $\sigma_r = 100$ MPa and 0.14 for $\sigma_r = 80$ MPa.
Longitudinal position (1–12) in the original bridge	No trend
Transverse position (A or B) in the original bridge	No trend

A question one could ask is whether the shear forces at fatigue loading are larger than the force transferred by friction due to the clamping force of the rivets, i.e. whether the shear forces produce friction slip or not? A rough calculation before the testing (set out below) suggested that this was not the case.

Clamping force N (assuming the tension in the rivet to be 70% of the yield stress, cf. Table 3.1):

$\sigma_y \approx 300$ MPa

$$\Rightarrow N \approx 0.70 \cdot 300 \cdot 10^3 \cdot 314 \cdot 10^{-6} = 65.9 \; kN \quad (6.1)$$

$A_{rivet} = \dfrac{\pi \cdot 20^2}{4} = 314 \; mm^2$

Friction force F:

$\mu \approx 0.33$

$N = 65.9$

$$\Rightarrow F \approx 0.33 \cdot 65.9 = 21.7 \; kN \quad (6.2)$$

Shear force V:

$P_{max} = 318.7 \; kN$ (Table 6.1)
$S = 19.2 \cdot 10^5 \; mm^3$
$I = 2.86 \cdot 10^9 \; mm^4 \Rightarrow V = \dfrac{318.7 \cdot 19.2 \cdot 10^{-4} \cdot 120 \cdot 10^{-3}}{2.86 \cdot 10^{-3} \cdot 2} = 12.8 \; kN \quad (6.3)$
$c = 120 \; mm$
Two shear planes

$\underline{V < F \quad OK!}$

The great differences in the fatigue lives that were found in the testing (especially for the stress range of 80 MPa – see Table 6.2) can be explained – ignoring the normally observed great scatter in fatigue test results – by the fact that the stringers had been in service for almost 100 years. If one assumes a certain amount of fatigue damage already accumulated in the stringers, the relative difference does not become so great.

122 Fatigue Life of Riveted Steel Bridges

Table 6.5 A comparison between fatigue *fracture* lives (number of load cycles N) given in the structural steel design code for riveted connections (C=71/category D) and the average fatigue lives found in the tests.

Stress range σ_r (MPa)	Structural steel design code		Fatigue life according to the tests (average)
	Design value[1]	Fracture value[2]	
80	1 398 090	3 512 002	4 952 237
100	715 822	1 798 145	1 950 007

1) The characteristic value of N.
2) The mean value of N.

Consider this hypothetical example (assuming an already accumulated fatigue damage of $5 \cdot 10^6$ cycles of the same stress range as in the test):

Stringer	σ_r	ΔN_{test}	$\Delta N_{accum.}$	ΔN_{tot}
3B	80 MPa	$2.4 \cdot 10^6$	$5 \cdot 10^6$	$7.4 \cdot 10^6$
5B	80 MPa	$6.5 \cdot 10^6$	$5 \cdot 10^6$	$11.5 \cdot 10^6$

Irrespective of whether there is fatigue damage accumulated in the stringers, the fatigue lives according to the test results are high. In comparison to the fatigue design curve for riveted connections in the structural steel codes (C=71/category D), the fracture values are well above the curve (cf. Fig. 6.18). However, it is not quite correct to compare fracture values in tests with a design curve. In an attempt to compare the fatigue lives of the stringers in this test with the "fracture lives" at the same stress range level in the codes, a "statistical transformation" of the design values is made here, see Table 6.5.

The conclusions from a comparison of the fatigue (fracture) lives given in Table 6.5 are as follows (cf. also Table 6.2):

- The structural steel design code is underestimating the fatigue life of riveted connections. At $\sigma_r = 100$ MPa there is a fairly good agreement between the predicted fatigue life according to the structural steel design code and the actual test result (only 8% underestimation). At $\sigma_r = 80$ MPa, however, the predicted fatigue life given in the codes underestimates by 29% the actual fatigue life as determined by the test results.
- The fatigue life of four of the six stringers tested (see Table 6.2) is well above the predicted value according to the structural steel design code.
- The test results show the remarkably long service life of the stringers (almost 100 years). One thing is absolutely sure, if the virgin stringers had been fatigue tested back in 1896 when the bridge was built, their fatigue life would have been found to be even greater than that shown by these test results.

Although the stringers had already been subjected to almost 100 years of service loading (and environmental effects), they were tested in the condition that they were in when they arrived in the laboratory. No sand blasting or similar treatment was carried out that might have made it easier to inspect the stringers for fatigue cracks during loading. The idea was to simulate "real" conditions as far as possible. The observations made during the tests in the laboratory must be "transferable" to the inspection routines in field. Other observations that were made during and after the tests (in addition to those already made in this chapter) are:

- The fatigue crack was always initiated from a rivet hole in the bottom (tension) flange.
- The fatigue propagation rate was fast at the final stages of cracking in the bottom flange – however, no unstable crack growth could be detected. The fracture surface was fibrous (not crystalline).
- A fine powder due to fretting between interfacing plates could sometimes be observed at or near the final crack position.
- In the early part of fatigue cracking, say during crack propagation in the first "initiated" L-profile, the crack was almost invisible to the naked eye (as small as "sewing thread" anyway). When the test was temporarily stopped and the stringer unloaded, the crack completely closed, making it impossible to detect. Regardless of whether the crack was detectable during loading, there was an increase in vertical deflection due to a loss of bending stiffness in the cracked region.
- No corrosion at the cracked rivet hole edges was detected.
- The rivet holes were drilled.
- There was an obvious need to intensify the inspection of the test girders after the initiation of a fatigue crack. Despite prolonged fracture behaviour due to inherent structural redundancy, the major part of each fatigue life consisted of crack initiation, i.e. the forming of a crack. The crack propagation time was estimated to be about 5–10% of total fatigue life.
- After the complete fracture of the lower tension flange (i.e. both L-profiles severed), the stringer was able to carry a substantial static load, see Fig. 6.24.

6.4.2 Tension tests

As soon as the girders arrived at the laboratory, steel samples were taken from the free ends of L-profiles in the way described earlier in Chapter 5. The upper flange was chosen for this purpose, due to the fact that the lower flange needed to be intact for fatigue testing. For the tension test, 24 specimens (sized $8 \times 10 \times 125$ mm^3) were machined, and these specimens were taken from six different stringers. The specimens were taken from the two ends of each L-profile in the upper (compression) flange, giving a total number of 24 specimens ($6 \times 2 \times 2$). Of these 24 specimens, 12 were chosen for an "ageing test", i.e. a tension test after prior cold-work deformation (strain 4–13%) and accelerated heat treatment (one hour at 200°C). A possible ageing effect is supposed to be fully developed after such treatment, cf. section 4.4.

The statistical calculations, which are made at the bottom of Table 6.6 showing the results of the tension test, comprise the average value, the standard deviation and the characteristic value. The latter is the lower 5%-fractile (i.e. 95% of the specimens have a higher yield stress) with a 75% confidence level.

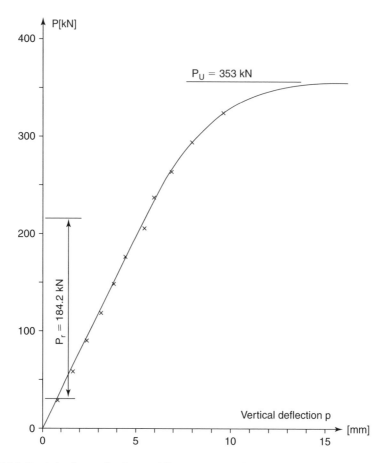

Fig. 6.24 Static load test *after* fatigue failure in stringer 3B ($\sigma_r = 80$ MPa). The vertical deflection is exclusive of the remaining (plastic) deformations after the completed fatigue test. The elastic vertical deflection during fatigue testing of the uncracked girder was 2.6 mm. The test can also be seen as a simple fracture toughness test. The initial fracture (read, defect) due to the fatigue failure was that both L-profiles in the lower tension flange were broken and there was a 370 mm vertical crack up into the web plate. As can be seen, the behaviour at maximum load is rather ductile.

The conclusions that can be drawn from the results of the standard tension test and the "ageing test" are:

- The mechanical properties are those of an ordinary mild steel of a lower strength class (but fully adequate).
- In comparison with the results for the other riveted bridges investiagted in Chapter 5, the characteristic yield stress level (256 MPa) is fairly high. The scatter is low, as

Table 6.6 Results of the tension tests on the flange material from the riveted girders.

Stringer number	Thickness t [mm]	Width b [mm]	Yield point		Tensile strengh		Elongation		Cold-working
			R_{eL} [MPa]	R'_{eL} [MPa]	R_m [MPa]	R'_m [MPa]	A_5 [%]	A'_5 [%]	$\Delta\varepsilon$ [%]
2A-L[1]	8.15	9.98	279.1	–	438.9	–	26.2	–	–
2A-L	8.06	9.91	290.5	(434.4)[2]	–	(490.8)	–	(29.2)	4.0
2A-R	8.16	10.02	266.6	–	422.0	–	32.2	–	–
2A-R	8.20	9.96	296.3	(468.9)	–	(481.2)	–	(26.4)	13.0
5A-L	8.16	10.04	266.1	–	421.1	–	39.0	–	–
5A-L	8.20	9.96	280.4	(448.1)	–	(476.3)	–	(29.2)	6.6
5A-R	8.10	10.02	284.6	–	441.1	–	34.8	–	–
5A-R	8.13	9.98	273.6	(449.9)	–	(464.6)	–	(31.0)	8.8
5B-L	8.13	9.96	274.2	–	443.3	–	35.2	–	–
5B-L	8.14	9.83	302.7	(479.0)	–	(492.1)	–	(26.8)	8.0
5B-R	8.15	10.03	260.6	–	424.5	–	33.8	–	–
5B-R	8.10	10.01	276.6	(457.6)	–	(479.8)	–	(27.2)	7.6
7A-L	8.06	9.96	250.4	–	382.4	–	38.0	–	–
7A-L	8.10	10.00	267.6	(429.6)	–	(450.6)	–	(32.0)	8.4
7A-R	8.15	9.94	274.0	–	417.2	–	36.4	–	–
7A-R	8.16	9.99	277.2	(461.2)	–	(487.0)	–	(29.4)	8.0
10B-L	8.07	9.99	281.6	–	434.1	–	35.4	–	–
10B-L	8.11	10.06	289.3	(467.0)	–	(492.7)	–	(28.2)	7.4
10B-R	8.06	9.73	281.8	–	431.0	–	30.6	–	–
10B-R	8.10	10.00	281.5	(448.1)	–	(470.4)	–	(30.2)	8.8
11A-L	8.07	8.99	270.4	–	419.3	–	40.4	–	–
11A-L	8.18	10.02	287.9	(457.5)	–	(481.3)	–	(30.8)	9.4
11A-R	8.07	9.98	279.4	–	430.8	–	40.0	–	–
11A-R	8.12	9.98	277.6	(418.3)	–	(450.4)	–	(34.6)	7.4
Average:			277.9	(415.6)	425.5	(476.4)	35.2	(29.6)	8.1
Standard deviation:			11.4	(17.5)	16.2	(14.8)	4.1	(2.3)	
Characteristic value:			256.1	(415.2)	391.8	(445.6)	26.7	(24.8)	

1) L = top left L-profile, R = top right L-profile.
2) Values in parentheses refer to results of tension tests after cold-working and ageing (i.e. heat treatment).

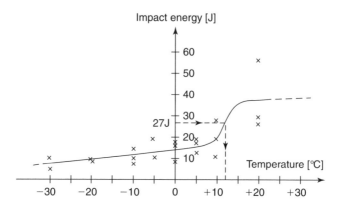

Fig. 6.25 The results of the impact notch tests (Charpy-V).

expressed by the standard deviation, giving a comparatively high characteristic yield stress value. There is also reason to believe that a certain ageing effect during the almost 100 years of service loading has increased the yield point and lowered the fracture elongation compared with these properties in the virgin steel material. Unfortunately there seems to be no record of any tensile tests on the virgin material used to construct this bridge to verify this hypothesis.

– The difference in the results from different stringers is negligible.
– The difference in the results from the top left and top right L-profiles in the same stringer is not more pronounced than the difference between specimens taken from the same L-profile.
– The ageing effect found from the tests used here was about the same as observed in previous studies (see Chapter 5):

- 63% increase in yield stress.
- 12% increase in tensile strength.
- 16% decrease in fracture elongation.

A comment must be made about this large increase in yield stress: a major part is due to strain hardening effects during the tension test.

– No difference can be observed between different degrees of cold-work deformation for the specimens in the ageing test.

6.4.3 Impact testing

A Charpy-V impact notch test series was performed on the flange steel from eleven stringers. The specimens (having a size $5 \times 10 \times 55$ mm^3) were tested for temperatures in the range +20 to −30°C, see Fig. 6.25.

The conclusions that can be drawn from the results are:

– The transition temperature is much too high to be accepted (according to the adopted impact toughness requirements) for a railway bridge located in the northern parts of Sweden (where temperatures can fall to −30°C or lower in winter).

Table 6.7 Results from the chemical analysis.

	2A [%]	6B [%]	11B [%]
Carbon	0.132	0.180	0.184
Sulphur	0.022	0.026	0.025
Phosphorus	0.08	0.06	0.08
Silicon	0.03	0.03	0.03
Manganese	0.41	0.38	0.40
Chromium	<0.01	<0.01	<0.01
Molybdenum	<0.01	<0.01	<0.01
Nickel	<0.01	<0.01	<0.01
Copper	0.04	0.03	0.03
Titanium	<0.01	<0.01	<0.01
Vanadium	<0.01	<0.01	<0.01
Cobalt	0.01	0.01	0.01
Nitrogen	0.0023	0.0025	0.0028

– The transition temperature is one of the highest (read, worst) ever reported for riveted railway bridges.

6.4.4 Chemical analysis

Three stringers were chosen for a chemical analysis of the steel, namely stringers 2A, 6B and 11B. The results of the analysis are shown in Table 6.7.

The conclusions that can be drawn from the results are:

– All three stringers show an alloy content normal for ordinary carbon steel.
– The nitrogen content is low, indicating steel that is not prone to ageing, such as Martin steel.
– The silicon content is low, indicating rimmed steel. However, the process of "killing" steel was not known at that time of the bridge's construction.

6.4.5 Clamping force

Nine rivets in the midspan region from stringer 3B were chosen in order to determine the clamping forces in the riveted connections. Six rivets were taken from the upper (compression) flange and three were taken from the lower (tension) flange. The elongation, which is forced upon the rivets by the connected plates (which resist the thermal contraction of the rivets after riveting), was relieved by machining away the outer parts of one of the rivet heads. This, at least, was the original idea, but it turned out that the misalignment of the rivet hole in the connected plates resisted most of the expected contraction. After experiencing bending deformations (which definitively was not the intention) when pressing out the first three rivets with hydraulic jacks (a pressure force of as much as 80 kN was required), the "dismantling process" was radically changed. For the last six rivets, the connected plates were carefully machined away in order to completely release the rivets without forcing any additional deformations upon them.

Table 6.8 Initial length L and the "recovery length" L' for the rivets that were chosen from stringer 3B for the determination of clamping forces.

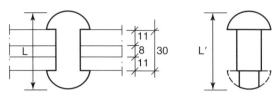

Rivet number	Position in the stringer	Initial length L[1] (mm)	Recovery length L'[1] (mm)	Contraction ΔL (mm)
1	Lower flange	54.858	—[2]	—
2	Lower flange	54.863	—[2]	—
3	Lower flange	54.876	—[2]	—
4	Upper flange	55.252	55.222	−0.030
5	Upper flange	55.262	55.231	−0.031
6	Upper flange	55.298	55.262	−0.036
7	Upper flange	55.200	55.179	−0.021
8	Upper flange	55.212	55.191	−0.021
9	Upper flange	55.225	55.192	−0.033

1) The average of five different measurements. The differences between the separate values and the average value were at the most ±0.003 mm.
2) Impossible to determine due to the bending deformations (explained in the text).

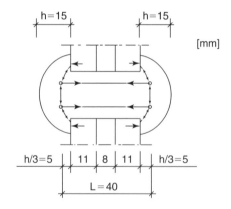

Fig. 6.26 Effective deformation length in the rivets.

The "effective deformation length" was determined to 40 mm, i.e. 5 mm up into the rivet heads, see Fig. 6.26.

Thus the strain $\Delta L/L$ could be determined and the clamping stress in each rivet shank was calculated, see Table 6.9.

A comment must be made here: the calculated clamping stress is highly dependent on the chosen value of the deformation length, cf. Fig 6.26. A choice of a shorter deformation length would result in a larger value for the clamping stress. Due to the

Table 6.9 Clamping stresses in the rivets.

Rivet number	Clamping stress σ_{cl} (MPa)
4	158
5	163
6	189
7	110
8	110
9	173
Average:	151
Standard deviation:	33

Table 6.10 The results from the tension test of the rivets.

Rivet number	Yield stress σ_y (MPa)	Tensile strength σ_B (MPa)	Elongation A (%)
1[1)]	–	–	–
2[2)]	(311)	(464)	(20.6)
3[2)]	(322)	(472)	(31.3)
4	377	458	23.1
5	328	425	29.4
6	373	509	24.0
7	353	509	22.3
8	324	454	28.9
9	376	442	23.0
Average:	355	466	25.1
Standard deviation:	24	34	3.2

1) Were excluded due to excessive bending deformations.
2) Not included in the average values due to prior bending deformations.

already short rivet shank length, the choice of a correct deformation zone in the rivet heads is highly important for the accuracy of the results.

In order to determine the "clamping ratio", i.e. the relation between the clamping stress σ_{cl} and the yield stress σ_y of the rivets, a tension test of the rivet material was performed, see Tables 6.10–11.

This type of investigation of the clamping force of old riveted structural members has, to the author's knowledge, not been performed before, and there are several circumstances that explain why the clamping ratio of the rivets tested here is lower than that normally reported (i.e. 70–90% of the yield stress):

– This investigation is about the *residual* clamping force after almost 100 years of service loading and after completed fatigue loading. Most other investigations concerned the *initial* clamping force, i.e. a direct dismantling immediately after riveting has been completed.
– The clamping stresses here are expressed in relation to the yield stress of the actual rivet, and not to the "nail" (i.e. the rivet before it is put in place) and also not to

Table 6.11 Clamping ratio of the different rivets tested.

Rivet number	Clamping stress σ_{cl} (MPa)	Yield stress σ_y (MPa)	Clamping ratio σ_{cl}/σ_y
1	–	–	–
2	–	(311)	–
3	–	(322)	–
4	158	377	0.42
5	163	328	0.50
6	189	373	0.51
7	110	353	0.31
8	110	324	0.34
9	173	376	0.46
Average:			0.42
Standard deviation:			0.08

the original rivet rod material. Both these operations, i.e. manufacturing a rivet "nail" out of a steel rod and putting the rivet into place, hardens the steel. If the clamping ratio was to be based on the yield stress of the original rod material, the relationship would be very much higher.

– The clamping force of the rivets in the upper flange can be assumed to be lower than the rivets in the lower flange. The upper rivets have, over the years, sustained a substantial amount of vertical shear loading from the train wheels. One can therefore assume that these rivets have a somewhat reduced clamping force due to effects like the fretting of the connected plate surfaces.
– The grip length is only 30 mm for these rivets, and in comparison to other riveted connections, this length is very small. As the grip length is increased, the clamping force increases.

6.4.6 Fatigue damage accumulation

In an attempt to estimate the fatigue damage that had already been accumulated in the stringers during the service period 1896–1993, a fatigue damage accumulation analysis was conducted by taking the actual loading history into consideration. The route from Vännäs to Umeå/Holmsund (the shipping port) has for many years been operated by heavy freight train traffic (with a maximum axle loading of 22.5 tons) carrying mainly pulp and timber. From the available train statistics it was possible to determine the loading history, see Fig. 6.27.

Using the method described in Chapter 5, this actual past loading data was then transformed into a number of "equivalent freight train passages" with the maximum allowed axle weight valid in 1993 (i.e. 22.5 tons). The "design train" for this particular railway line is freight wagon class D2 with two bogie-axles and a total length of 14.05 m for each wagon, see Fig. 6.28.

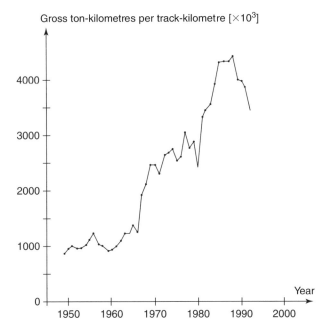

Fig. 6.27 Loading history during 1949–1992 for the railway bridge over the Vindelälven at Vännäsby.

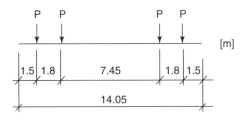

Fig. 6.28 Freight train class D2 ($P = 22.5$ tons).

The maximum stress range generated by this particular freight wagon when passing over (the influence length of) the stringers is equal to the maximum stress of one bogie-axle, see Fig. 6.29.

The gross ton-kilometres per track-kilometre are now transformed into a number of "equivalent bogie-axle passages" due to the fact that each such pair of axles can generate the maximum stress range by itself, see Table 6.12.

The predicted lifespan for these riveted stringers according to the structural steel design code is then derived (using $C = 71$ and without making any special safety factor considerations). There is also a "secondary" smaller stress range for each passage of

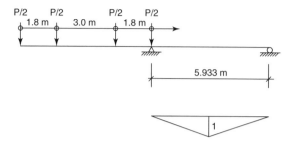

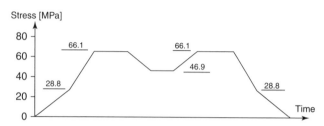

Fig. 6.29 Stress variation at the midspan of the stringer for the passage of a bogie-axle *pair* of freight wagon D2, assuming a simple support.

a bogie-axle pair, cf. Fig. 6.29, but this stress range of 19.2 MPa (66.1–46.9) is below the fatigue limit and therefore too small to generate any fatigue damage.

$$N = \frac{2 \cdot 10^6}{\left(\frac{\sigma_r}{C}\right)^3} = \frac{2 \cdot 10^6}{\left(\frac{66.1}{71}\right)^3} = 2{,}478{,}567 \tag{6.4}$$

The fatigue damage accumulation factor D is:

$$D = \frac{n}{N} = \frac{2{,}887{,}376 \cdot 0.5}{2{,}478{,}567} = 0.58 \quad (<1) \tag{6.5}$$

The coefficient 0.5 is applied because although there are 2,887,376 passages of *one* bogie-axle in total (cf. Table 6.12), the stress range is computed for a *pair* of bogie-axles (cf. Fig. 6.29).

This fatigue damage accumulation analysis suggests that, assuming an "equivalent number of freight train passages" (read, bogie-axle passages), the theoretical fatigue "fracture" life is longer than the actual number of years the bridge has been in service ($D = 0.58 < 1$). The remaining fatigue (fracture) life, assuming an annual increase in freight train transport of 5%, is 16 years (i.e. when $D = 1$ is reached). This method errs perhaps too much on the *safe side* (considering the results of the fatigue testing), but it is a very simple tool in estimating the fatigue damage accumulation and the remaining fatigue life of railway bridges.

Table 6.12 Transformation of the loading statistics for each year into a number of "equivalent bogie-axle passages" of 2 × 22.5 tons (assumed values within parenthesis).

Year	Gross ton-kilometres per track-kilometre	Equivalent bogie-axle passages	Year	Gross ton-kilometres per track-kilometre	Equivalent bogie-axle passages
1896	(450·10³)	10 000	46	(600")	13 333
97	(")	"	47	(700")	15 555
98	(")	"	48	(800")	17 777
99	(")	"	49	866"	19 244
1900	(")	"	1950	935"	20 777
01	(")	"	51	1000"	22 222
02	(")	"	52	969"	21 533
03	(")	"	53	964"	21 422
04	(")	"	54	1071"	23 800
05	(")	"	55	1106"	24 577
06	(")	"	56	1227"	27 266
07	(")	"	57	1030"	22 888
08	(")	"	58	978"	21 733
09	(")	"	59	922"	20 488
1910	(")	"	1960	953"	21 177
11	(")	"	61	983"	21 844
12	(")	"	62	1075"	23 888
13	(")	"	63	1222"	27 155
14	(")	"	64	1219"	27 088
15	(")	"	65	1353"	30 066
16	(")	"	66	1247"	27 711
17	(")	"	67	1922"	42 711
18	(")	"	68	2137"	47 488
19	(")	"	69	2471"	54 911
1920	(")	"	1970	2468"	54 844
21	(")	"	71	2321"	51 577
22	(")	"	72	2645"	58 777
23	(")	"	73	2691"	59 800
24	(")	"	74	2750"	61 111
25	(")	"	75	2528"	56 177
26	(")	"	76	2619"	58 200
27	(")	"	77	3051"	67 800
28	(")	"	78	2811"	62 466
29	(")	"	79	2887"	64 155
1930	(")	"	1980	(2424")	53 866
31	(")	"	81	3321"	73 800
32	(")	"	82	3460"	76 888
33	(")	"	83	3599"	79 977
34	(")	"	84	3912"	86 933
35	(")	"	85	4304"	95 644
36	(")	"	86	4341"	96 466
37	(")	"	87	4347"	96 600
38	(")	"	88	4432"	98 488
39	(")	"	89	4007"	89 044
1940	(")	"	1990	3954"	87 866
41	(")	"	91	3863"	85 844
42	(")	"	92	3449"	76 644
43	(")	"	93	(3449")	76 644 Σ2,887,376
44	(")	"			
45	(500·10³)	11 111			

If we instead assume a more realistic maximum stress range value – say $\sigma_r = 40$ MPa – the result is quite different (the expression for the number of stress cycles differs from the expression used above because the Swedish code (in use in the early 1990s) has another curve slope for values of N above $5 \cdot 10^6$:

$$N' = \frac{2 \cdot 10^6}{\left(\frac{\sigma_r}{0.885 \cdot C}\right)^5} = \frac{2 \cdot 10^6}{\left(\frac{40}{0.885 \cdot 71}\right)^5} = 19,131,067 \qquad (6.6)$$

The fatigue damage accumulation factor D' for this case is:

$$D' = \frac{n}{N'} = \frac{2,887,376 \cdot 0.5}{19,131,067} = 0.08 (<<1) \qquad (6.7)$$

As can be seen, there is a large difference in the lifespan results depending on the choice of stress range. But, considering that a maximum stress range level of 40 MPa is more realistic and that most of the freight wagons (especially in the earlier years of the bridge's working life) only had two axles (and therefore generated smaller stress ranges in the stringers), one can assume that the fatigue damage accumulation factor D is *very* small, say $D << 0.1$.

6.5 Comparison with other investigations

The results from this full-scale fatigue test series of nine riveted stringers can be compared with the results from previous studies (cf. Chapter 2). Plotted on the same Wöhler diagram as given in section 2.3.10, i.e. the summary of all previous full-scale fatigue test results, it can be seen that the results from the full-scale fatigue tests conducted at the Department of Structural Engineering at Chalmers University of Technology agree very well with previous findings, see Fig. 6.30.

The principal findings and observations from the full-scale fatigue test compared with previous studies are as follows:

- The fatigue lives are well above the fatigue design curve for riveted details given in the codes ($C = 71$/category D). Previous studies on non-corroded riveted bridge members taken out of service all show similar results.
- No influence on the fatigue lives from differences in clamping forces was found, since no investigation of the clamping force for different stringers was made. However, Reemsnyder [18], and Baker and Kulak [22] showed that an increase in clamping force also results in an increase in fatigue life.
- No reducing effect on the fatigue life from the presence of corrosion damage could be observed. Several investigations show similar results – if the corrosion damage is not too severe, and if the rivet head protects the rivet hole edge from any influence of corrosion, then the fatigue life is the same as that of a riveted member without corrosion damage.
- The final fatigue failure of the cracked stringers exhibited non-brittle fracture behaviour. All the tests were performed at room temperature though. Fisher et al. [24] performed a reduced temperature test (the temperature was dropped to $-40°C$ or below at periodic intervals during fatigue testing), and found no sign of any unstable crack growth (i.e. brittle fracture).

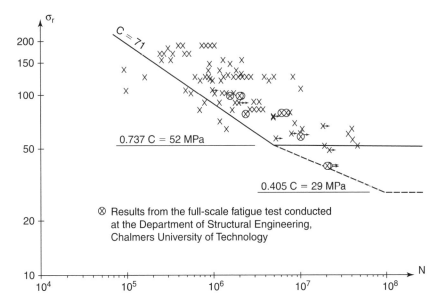

Fig. 6.30 The results from the full-scale fatigue tests conducted at the Department of Structural Engineering at Chalmers University of Technology (i.e. this study) and the results from previous studies compared with the fatigue design curve for riveted details given in the codes ($C = 71$/category D).

- No direct correlation between mean stress and fatigue life could be ascertained. Fisher et al. [24] let the mean stress vary between 56, 107 and 159 MPa during a fatigue test series. They were also not able to confirm any influence on the fatigue life from the mean stress level.
- Despite a thorough analysis of the loading history – no effect on the fatigue life from previous fatigue damage accumulation during service loading could be determined. Abe [25] performed a study on differences in stress range histories for different bridge member parts. The result was the same, no influence from fatigue damage accumulation could be determined.

6.6 Future investigations

In this full-scale fatigue test of nine stringers taken from the riveted railway bridge over the Vindelälven at Vännäsby, the main interest has simply been the remaining fatigue life. There are, of course, many other (closely related) subjects that are very important to study in future investigations. Here are the most important and relevant topics for further study (in no particular order):

- The full-scale fatigue test series must be continued with other stringers from this bridge in order to ascertain a statistically more satisfying result. After performing a sufficient number of tests for different stress range levels, it will be possible to determine a fatigue design curve for this kind of riveted full-scale specimen.

- The connection between stringers and floor beams has been shown to be critical over the years (a number of loose and faulty rivets together with fatigue cracks have been reported in several different bridges). A full-scale fatigue test programme on these joints is most important.
- More full-scale fatigue tests on stringers and floor beams are needed in order to determine their "defect fatigue resistance". Several different defects (such as a locally buckled flange plate or a sharp notch at the edge of the flange) can be made in beams in order to simulate hit damage. What is the reduction in the fatigue life that results from different types of damage? What extent (or size) of defect is required to "overlap" the effect from the rivet holes (i.e. which defect governs the fatigue life)?
- It is important to perform full-scale fracture mechanics tests of the stringers in order to determine their fracture behaviour. The testing must, however, be based on real parameters rather than standardized procedures. Stress level, strain rate, defects likely to occur and composite action are all very important parameters to consider more thoroughly. The results must be applicable to actual bridges still in service and not to some "hypothetical worst case".
- If the rivets in critical locations (e.g. flange-rivet connections for wind-bracing diagonals) are replaced with high-strength friction-grip bolts, what is then the effect on the fatigue life? Is it, for instance, possible to increase the remaining fatigue life by replacing a rivet after a crack has been detected?
- What is the effect on the remaining fatigue life of drilling a so-called "stop hole" at the tip of a propagating fatigue crack? In addition, is it possible to further increase the remaining fatigue life by adding a high-strength friction-grip bolt in the hole?
- What degree of accuracy in terms of crack detection is achieved by ultrasonic testing?
- Is it possible to determine any difference in results between spectrum loading and constant-amplitude loading for riveted members?
- How should a proper fatigue damage accumulation analysis be performed?
- What is the effect of ageing and temperature on fatigue life?
- What different reinforcement methods can be used to achieve a longer service life?

Chapter 7

Summary and Conclusions

This final chapter (in the 1994 thesis) is divided into three parts. First, there is a summary of the results from the full-scale fatigue test series of nine riveted stringers taken from the railway bridge over the Vindelälven at Vännäsby. Second, there is a summary of the results and findings from the field investigations of 15 riveted railway bridges, and third, some general conclusions are drawn from all the results. The questions given in section 1.2 ("Aim") are also answered.

Full-scale fatigue test series

- The fatigue lives of the nine stringers tested were in accordance with or above the fatigue design curve given in the structural steel codes for riveted details ($C=71$/category D), despite suffering almost 100 years of loading.
- The fatigue cracks always started and subsequently propagated in the lower (tension) flange.
- The fatigue crack initiations were, in five cases out of six, at a flange-rivet connection outside the midspan region. Only in one case did the fatigue crack emanate from a neck-rivet hole in the midspan region.
- A substantial number of loading cycles were required to propagate the fatigue crack from one L-profile to the other (on average, about 100,000 cycles). The riveted built-up composite I-beams (i.e. the stringers) thus showed an inherent structural redundancy.
- The fatigue cracks were difficult to observe at the beginning of the propagation period. They were almost invisible to the naked eye when the stringers were temporarily unloaded.
- The fractured stringers were capable of carrying a considerable static load (greater than the maximum load of the fatigue test) after completed cyclic loading.
- A tension test of the steel material showed that its mechanical properties were comparable to mild steel with a yield point higher than many other riveted bridge steels that have been tested.
- A marked ageing effect was found in the steel after cold deformation and subsequent heat treatment.
- The result of the Charpy-V impact notch test was a transition temperature of approximately $+12°C$. This temperature is much higher than the lowest service temperature.

- A chemical analysis confirmed that the bridge material is an ordinary carbon steel with low nitrogen content.
- The clamping ratio (i.e. rivet clamping stress in relation to the yield stress of the rivet) was found to be 0.42 on average.
- The fatigue damage accumulated in the stringers tested was found to be negligible.

Field investigations

- No major corrosion damage was found, despite maintenance being neglected in some cases.
- No loose or faulty rivets were detected.
- No visible cracks were found.
- The maximum stress in a bridge superstructure seldom exceeded 42 MPa when heavy freight trains were passing.
- There is a good correspondence between theoretical strain values and the actual strain response. The measured strain response is generally somewhat lower than the predicted response.
- The dynamic amplification factor was found to be of smaller magnitude than that usually assumed.
- The maximum deflection of the bridges is quite small, and much below that allowed by the railway bridge codes.
- Extensive ultrasonic crack controls were performed, and fatigue cracks were not detected at any rivet holes.
- Tension tests revealed that the bridge material is of ordinary mild steel quality (a mean yield stress of approximately 285 MPa on average and an elongation at rupture of about 30%).
- Ageing tests of the steels gave the following average results:
  - 50% increase in yield point.
  - 10% increase in tensile strength.
  - 13% decrease in elongation at rupture.
- A Charpy V-notch impact test showed that the transition temperature was well above the highest allowable temperature according to the service conditions.
- The nitrogen content did not exceed 0.007% in any material tested, and, together with a relatively low content of phosphorus, there is a good reason to assume that none of the bridges were constructed from Thomas steel.
- The grain sizes were between 30–60 μm.
- The static load-carrying capacity is in many cases sufficient to carry modern "design traffic" (i.e. train load UIC 71 according to the railway bridge codes). An increase in the highest allowable axle load for the bridges from 18 tons (original design) to 22.5–25 tons has been possible.
- A fairly accurate estimate of the remaining fatigue life has been possible through a thorough study of the loading histories. The bridges that have been investigated have a remaining fatigue life of 27 years or more.
- By a fracture mechanics analysis it was found that the critical crack length (a crack emanating from a rivet hole) can be assumed to be in the neighbourhood of 25 mm.

General conclusions

The aim of this dissertation was to show that there is a substantial remaining fatigue life in typical riveted railway bridges still in use at the time of the study (the early 1990s). Both the full-scale fatigue tests and the different field investigations clearly show that this is the case:

- The full-scale fatigue tests showed that the fatigue design curve for riveted details underestimates the fatigue life. However, this curve can be used as a safe estimate for the remaining fatigue life of old riveted railway bridges still in service.
- The field investigations showed that the maximum stress range seldom, if ever, exceeds the fatigue limit for riveted details. The fatigue damage accumulated in the bridges can therefore be assumed to be negligible.

In those cases where fatigue cracks have been found, it has been a result of *secondary constraints* (e.g. in the connection between stringers and floor beams). Secondary bending stresses are impossible to avoid and they can, in the worst case, initiate a fatigue crack in the overstressed region. But as soon as a crack has been formed, there is a "release" of the restraint, and the formerly overstressed region is "relaxed". A fatigue crack caused solely by a normal stress variation (i.e. the normal "design behaviour") has not, to the author's knowledge, ever been found in riveted railway bridges in Sweden.

Another question that is very relevant for the expected service life of a riveted railway bridge is what is the probability of a brittle fracture. The fracture toughness of the steel has been shown to be too low in respect to the generally accepted requirements regarding sufficient ductility. But, if so, why have there not been any brittle fractures? There are several possible answers to this question:

- Low stresses in general (including low dynamic amplification).
- The stresses have not been large enough to initiate fatigue cracks (and, subsequently, to propagate these cracks to a certain critical length).
- The major part of the fatigue life (say 90–95%) is the period before the initiation of a fatigue crack, and therefore the probability of a brittle fracture during normal service life is negligible.
- The strain rate is in general low (the fracture toughness increases as the strain rate decreases).
- The plate thicknesses are small (the fracture toughness requirements [that apply in the early 1990s] are based on thick plates).
- There is normally no other defect in the primary tension members other than the rivet holes, and this stress raiser is by no means large enough to be the single cause for the initiation of a brittle fracture.
- The clamping force is introducing a triaxial stress state at the rivet hole that prevents any fracture tendencies.
- A riveted built-up bridge member shows an inherent structural redundancy through "separate part composite action". A crack will, without exception, be stopped when passing from one member part to another.
- Large stress concentrations are unlikely to occur, since overstressed regions will be relieved by friction slip.

Taking all these factors together, it is the author's opinion that the fracture toughness of old riveted bridge steel can be assumed to be sufficient, despite failing to meet contemporary toughness requirements. And additionally, an eventual ageing effect, which reduces the fracture toughness further, can also be assumed to be negligible as long as the primary members are not being subject to any cold deformation (e.g. being hit and damaged).

Finally, the questions given in section 1.2 ("Aim") will be answered here:

What is the expected fatigue life of a riveted railway bridge member?
Is it possible to assume a very long (almost infinite) fatigue life for different riveted members subjected to stress ranges below the fatigue limit?
Given the stress levels generated by everyday traffic, one can – in principle – assume an almost infinite life. However, at the presence of secondary (bending) stresses, non-propagating fatigue cracks may develop.

What is the level of stress ranges expected for everyday train traffic?
At most, stress cycles with a maximum level of 42 MPa for typical bridge members.

How does temperature and the effect of ageing influence fatigue crack propagation?
The fatigue limit is increased when there is an effect of ageing and when the temperature is lowered (i.e. this is beneficial). The crack propagation rate is, however, unaffected once a crack has been formed.

How should one take different loading histories into consideration when carrying out a fatigue damage accumulation analysis?
By transferring the past loading into a number of "equivalent freight train passages" it is possible to calculate the accumulated fatigue damage.

How should a proper fatigue crack detection control be made?
By ultrasonic testing of the most overstressed regions, it is possible to detect fatigue cracks hidden behind the rivet heads.

In what parts and at what locations in the structure are the fatigue cracks to be expected?
The stringers are, due to the very short influence length, suffering more stress cycles for each train passage than other members, and they are therefore the most likely bridge members to experience fatigue cracking.

After how many stress cycles is a fatigue crack expected to become visible?
After about 90–95% of the fatigue life. Given different constant amplitude stress range values:

$$100 \text{ MPa} \quad N \geq 1{,}500{,}000 \text{ cycles}$$
$$80 \text{ MPa} \quad N \geq 4{,}500{,}000 \text{ cycles}$$
$$60 \text{ MPa} \quad N \geq 10{,}000{,}000 \text{ cycles}$$
$$40 \text{ MPa} \quad N \geq 20{,}000{,}000 \text{ cycles}$$

How are loose rivets affecting the fatigue life of different bridge members?
The fatigue life of a riveted joint is directly proportional to the magnitude of the clamping force. It is therefore of absolute importance that loose rivets are immediately replaced in order to maintain adequate fatigue resistance.

How should the probability of defects other than fatigue cracks be taken into consideration?
Through proper maintenance and inspection routines that safely detect major faults in time, it is possible to keep the structure "upgraded". Any defect (including fatigue cracks) that affects the load-carrying behaviour should not be allowed.

Is it possible to repair and strengthen these old riveted structures?
Definitively yes. If the idea is to add material to the existing structure, welding should be ruled out a joining method. The steel may very well be weldable, but, nevertheless, one should not weld on an old riveted steel bridge. The steel becomes even more brittle and the fatigue strength is lowered markedly.
The full-scale fatigue test in this dissertation indicates that replacing rivets in the flange-connections with high-strength friction-grip bolts would prolong the fatigue life of the original bridge.

What should a proper inspection manual for the inspection routines concerning fatigue cracks look like?
Clear and simple drawings showing the most probable locations in the bridge that could be experiencing fatigue cracking. In addition to the drawings, some explanatory text setting out the main reasons for the cracking.

The particular bridge that was chosen for the full-scale fatigue test series that was performed in this study must be said to have been replaced too soon based on the fatigue life results obtained from the investigation. If the results of the fatigue tests had been known in advance of the decision to replace the bridge, and if a thorough analysis of the structural load-carrying system had been performed (revealing the inadequate sway-bracing system, for example), the railway bridge over the Vindelälven at Vännäsby could have been used in service for many more years. These old riveted railway bridges, dating from the late nineteenth century and the first part of the twentieth century, must be especially cared for since they also represent a great cultural heritage!

Chapter 8

Modern Research

This chapter presents examples of research undertaken in the period 1995–2009 concerning *full-scale fatigue testing* of riveted bridge members, i.e. research conducted in the same field as my dissertation which was first published in October 1994.

In 1995 Adamson and Kulak performed a fatigue test series on six riveted stringers taken from a railway truss bridge that was built in 1911 [122]. The stringers were subjected to four-point bending using stress ranges in the interval of 62–73 MPa. All six girders were subjected to loading cycles, until fatigue cracking, which exceeded detail category $C=71$ (AASHTO category D). In fact, for two of the girders the test was completed without any fatigue cracking at all. The cracks were initiated in the lower tension flange, either from the rivet holes connecting the L-profiles to the web plate or at the position where horizontal gusset plates were attached.

Dibattista and Kulak used four tension diagonals (taken from the same railway bridge used by Adamson and Kulak [122]), which all were cut in two in order to perform a fatigue test series in 1995 on these specimens [123]. The specimens had the upper and lower gusset plate connections remaining (for the original connection to the upper and lower chord respectively), and they were subjected to axial cyclic tension with stress ranges in the interval of 67–73 MPa. It was found that the specimens that had the lower chord gusset plate connection experienced fatigue lives that *exceeded* detail category $C=71$ (AASHTO category D), and that the specimens that had the upper chord gusset plate connection experienced fatigue lives that were *below* detail category $C=71$ (AASHTO category D). All cracks were located in the shear splice end connection between the tension diagonal and the gusset plate.

The findings from full-scale fatigue tests on 20 riveted bridge girders – eight at variable amplitude loading and twelve at constant amplitude loading – were reported in 1995 by Zhou, Fisher and Sweeney [124]. For five of the girders – tested at a stress range of between 45–54 MPa for up to $100 \cdot 10^6$ loading cycles – no fatigue cracks were detected, but for the other girders the test resulted in 60 cracks.

Helmerich, Brandes and Herter carried out a fatigue test series on three different girder types in 1997 [127]. Nine girders in all – two mild steel riveted lattice girders (taken from a railway bridge that was built in 1902), three mild steel riveted built-up girders (taken from an old bridge) and four wrought-iron riveted built-up cross-girders (taken from a bridge built in 1890) – were tested by bending at stress ranges in the interval of 85–135 MPa. The fatigue lives were all observed to agree with detail category $C=71$ (AASHTO category D), and the fatigue cracks initiated from rivet holes, in gusset plates for the lattice girders and in cover plates or L-profiles for the built-up girders. It was also found that if a row of rivets were replaced by high-strength bolts in the cover plates for the wrought-iron cross-girders, the fatigue life increased considerably.

Three riveted built-up stringers, taken from the old railway bridge at Vännäsby, Sweden, were tested by Kadir in 1997 [128], two at 100 MPa and one at 97 MPa. The results were in accordance with or above detail category C = 71 (AASHTO category D).

In 2001 Crocetti presented the results from constant amplitude fatigue limit (CAFL) tests on five riveted built-up stringers (taken from the old railway bridge at Vännäsby) [132]. The girders were loaded cyclically at two stress range levels, 40 MPa and 60 MPa, for $10–20 \cdot 10^6$ loading cycles. In order to determine whether this cyclic loading had produced any fatigue damage, the girders were tested a second time. This second test was at a much higher loading level, about 100 MPa. The tests continued until the girders failed by fatigue cracking, and the test results were then compared with other girders that previously had been tested at the department at the same high stress range level. It was found that the girders first tested by Crocetti at a stress range level of 40 MPa produced results similar to those obtained from girders previously tested only at 100 MPa, proving the fact that the loading at the lower stress range level had not produced any fatigue damage. However, for the girders that first were tested at 60 MPa and then at about 100 MPa, there was a slight deviation from the results obtained from the girders previously tested at only 100 MPa, proving that a stress range level of 60 MPa is just above an assumed CAFL. Crocetti proposed that the CAFL given by detail category C = 71 (which is 52 MPa) is a good estimate.

In 2001 Xie, Bressant, Chapman and Hobbs presented the results from a fatigue test series on three riveted cross-girders – 203 mm deep-rolled sections reinforced by 15 mm thick cover plates riveted to both flanges – taken from an old steel bridge (built in the 1880s) on the London Underground [133]. The girders were tested by four-point bending under variable amplitude loading and it was found that the fatigue lives were considerably greater than that given in BS 5400 category B.

Al-Emrani performed a fatigue test series in 2002 on three riveted built-up stringers (taken from the old railway bridge at Vännäsby) – tested by four-point-bending at a stress range level of 60–100 MPa – to study the effect of drilling stop holes at the tip of the propagating fatigue cracks [134]. It was found that the arrest of the propagation was considerable, delaying the renewed propagation by up to 215,000 loading cycles.

Al-Emrani also carried out a huge full-scale fatigue test study of secondary-effect cracking at stringer-to-floor-beam connections, using other parts from the old railway bridge at Vännäsby. These consisted of four stringers connected to three floor beams, all still connected together as in the original bridge. Three "packages" like this were tested in order to study the initiation and possible propagation of secondary cracks at L-profile shear connections between longitudinal stringers and transverse floor beams. All parameters that influence the fatigue behaviour of stringer-to-floor-beam connections were also studied in detail in a very thorough and extensive FE-analysis, where the results from the full-scale fatigue tests were simulated exactly.

In 2006 Matar and Greiner carried out a fatigue test series on four built-up riveted stringers (two different cross-sections) taken from two old bridges, built in 1903 and 1913. The girders were tested by three-point bending and the fatigue lives exceeded category ECCS 90. In addition, two more stringers were tested statically in order to study the clamping force of the rivets and the load/slip behaviour of the connections.

In 2009 Pipinato, Pellegrino, Bursi and Modena carried out a full-scale fatigue test series on a 12.4 m long open-deck riveted railway girder bridge, which was built in 1918 and taken out of service in 2006 [146]. The critical details that were studied were the transverse shear diaphragms, which connected the four longitudinal girders every 1.0 metres. In the original bridge the shear diaphragms carried the rails, and were consequently also carrying the pulsating load from the hydraulic jacks during the fatigue test series. The rivets connecting the diaphragms were subjected to a shear stress range of between 265–353 MPa, and it was found that the fatigue strength given by shear Category 100 of the Eurocode (equivalent to non-preloaded bolted splices) is on the safe side when estimating the fatigue lives of these riveted shear splices.

Examples

Example 1

Check if this riveted double-lap joint is able to carry a cyclic loading of $P_r = 75$ kN for a total of $10 \cdot 10^6$ loading cycles.

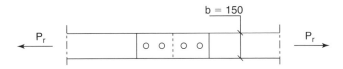

The joint is subject to periodic inspection and maintenance, and it is an accessible joint detail and a non "fail-safe" component.

$$\Delta \sigma = \frac{P_r}{b \cdot t} = \frac{75 \cdot 10^{-3}}{150 \cdot 14 \cdot 10^{-6}} = 35.7 \, MPa$$

$\Delta \sigma_D = 52 \, MPa$ (Detail category 71)

$$\gamma_{Ff} \cdot \Delta \sigma \leq \frac{\Delta \sigma_D}{\gamma_{Mf}} \Rightarrow 1.0 \cdot 35.7 \leq \frac{52}{1.25} \Rightarrow 35.7 < 41.6 \text{ OK (i.e. infinite life)}$$

The fatigue crack propagation scenario that could be expected is the forming of two cracks on either side of a rivet hole in the central plate (this could be any of the four rivet holes), where 1 is the crack initiation point (i.e. where the stress concentration effect is the largest) and 2 is the crack propagation direction (perpendicular to the applied normal stress).

However, for the case of large number of cycles, there could be a gradual loss of clamping force (due to fretting of the interfacing plates), resulting in localized bending

148 Examples

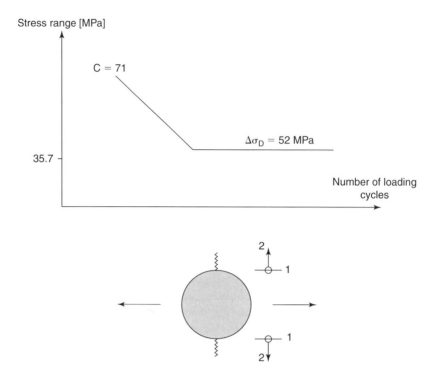

of the rivet head (due to the ovalization of the rivet hole). Under these circumstances a *single* fatigue crack is formed underneath one of the rivet heads (this could be any of the four) leading eventually to the loss of the head:

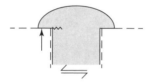

Example 2

In a laboratory test, two symmetrically positioned hydraulic jacks are each producing a pulsating load P (in four-point bending) on a riveted girder that has been dismantled and taken out of service from a railway bridge.

The pulsating load P is varied as follows: $P_{min} = 30\,\text{kN}$ and $P_{max} = 100\,\text{kN}$, in constant amplitude loading. How many loading cycles can be expected if we assume that no fatigue damage has been accumulated during the in-service loading?

$$I = \frac{8 \cdot 550^3}{12} + 4 \cdot 940 \cdot \left(\frac{550}{2} - 19.7\right)^2 = 3.56 \cdot 10^8 \, mm^4$$

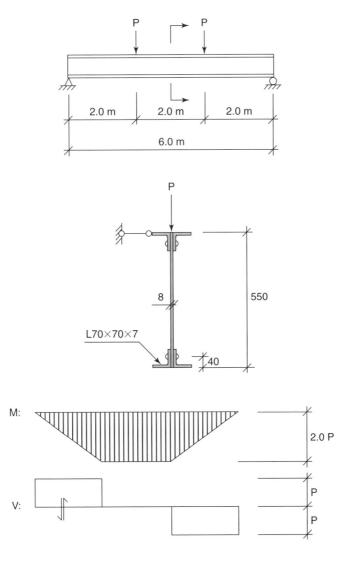

Stress range at the position of the lower rivet (on the tension side, where the fatigue crack will occur):

$$\Delta\sigma = \frac{2.0 \cdot (100-30) \cdot 10^{-3}}{3.56 \cdot 10^{-4}} \cdot \left(\frac{550}{2} - 40\right) \cdot 10^{-3} = 92.4 \, MPa$$

$$N = 5 \cdot 10^6 \left[\frac{\Delta\sigma_D}{\Delta\sigma}\right]^3 = 5 \cdot 10^6 \left[\frac{52}{92.4}\right]^3 = 8.9 \cdot 10^5$$

(no partial coefficients as this is analysis, not design)

150 Examples

Alernatively:

$$\log N = 11.851 - 3 \cdot \log 92.4 \Rightarrow N = 8.9 \cdot 10^5$$

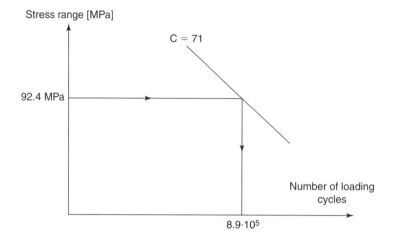

Example 3

This is based on the same test-up as in example 2, however now under the assumption that the fatigue loading using $P_{max} = 100\,\text{kN}$ and $P_{min} = 30\,\text{kN}$ is interrupted after 300,000 cycles, and that the test is continued using $P_{min} = 20\,\text{kN}$ and $P_{max} = 50\,\text{kN}$. How many loading cycles can be expected using this lower loading level?

$$P_{r2} = 50 - 20 = 30\,kN \quad \Delta\sigma_2 = \frac{30}{70} \cdot 92.4 = 39.6\,MPa\,(\Delta\sigma_L < \Delta\sigma_2 < \Delta\sigma_D)$$

$$\Delta\sigma_L = 29\,MPa \quad \Delta\sigma_D = 52\,MPa$$

This lower loading level does contribute to fatigue damage even though it is below the constant amplitude fatigue limit (cf. the contrary finding in example 1). In contrast to constant amplitude loading, stress ranges in spectrum loading (variable amplitude loading), in between the cut-off limit and the constant amplitude fatigue limit, contribute to fatigue damage, as long as the maximum stress loading in the spectrum exceeds the constant amplitude fatigue limit.

$$N_2 = 5 \cdot 10^6 \cdot \left[\frac{52}{39.6}\right]^5 = 19.5 \cdot 10^6$$

$$D = \Sigma \frac{n}{N} \leq 1 \quad \frac{300 \cdot 10^3}{8.9 \cdot 10^5} + \frac{n_2}{19.5 \cdot 10^6} \leq 1 \Rightarrow n_2 \leq 12.9 \cdot 10^6$$

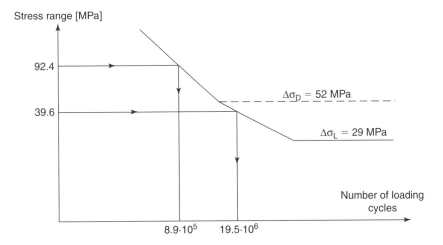

Example 4

After more than 70 years of service life, a continuous riveted truss bridge was accidentally hit by a boat passing underneath. Local damage was observed in the lower chord in the midspan of the left outer span. The damage was identified as a cut-like notch at the edge of the lower angle flange (through all the thickness), and the notch effect was judged equivalent to detail category 56. What is the maximum allowed axle weight P (in a group of three loads representing the heaviest vehicle passing the bridge), if the required additional service life is 15 years (after which the bridge is to be replaced by a new one)? Assume that no fatigue damage has been accumulated in the bridge prior to the collision. Assume the heavy vehicle (a lorry) crosses the bridge 10 times a day (both directions taken together) all year round, and can be assumed to be the only vehicle contributing to fatigue damage in this new situation. In order to calculate the stress in the damaged chord, we use the influence line (for a single load)

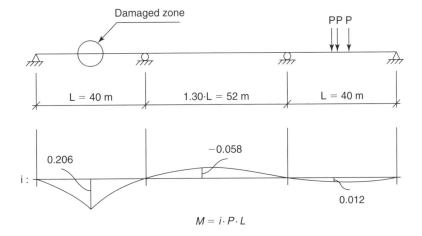

for a continuous girder (see the figure above), knowing that we have to transform the bending moment into a normal force in the chord. The position in the transverse direction of the axle loading is shown in the cross-section below. In the longitudinal direction the three axle loadings, being so close, can be seen as being just one load 3P. Regarding the safety factor γ_{Mf} it can, on the safe side, be chosen as being 1.35, equivalent to a non "fail-safe" component in combination with poor accessibility, even though it can be assumed that the damaged zone is kept under repeated close visual examination.

Elevation of the bridge and the influence line:
Transverse direction:

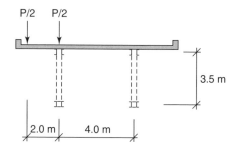

Longitudinal direction:

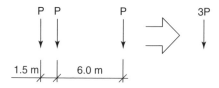

Truss elevation:

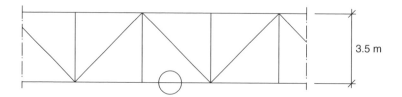

The lower chord:
Moment equilibrium around reaction force R_B:

$$R_A \cdot 4 - \frac{P}{2} \cdot (4+6) = 0 \Rightarrow R_A = 1.25 \cdot P$$

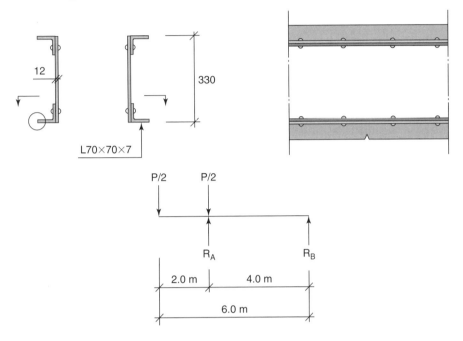

$$M_r = (0.206 + 0.058) \cdot (3 \cdot 1.25 \cdot P) \cdot 40 = 39.6 \cdot P$$

$$N_r = \frac{M_r}{h} = \frac{39.6 \cdot P}{3.5} = 11.3 \cdot P$$

$$A = 2 \cdot 12 \cdot 330 + 4 \cdot 940 = 11680 \, mm^2$$

$$\Delta\sigma = \frac{N_r}{A} = \frac{11.3 \cdot P \cdot 10^{-3}}{11680 \cdot 10^{-6}} = 0.967 \cdot P$$

$$\frac{\Delta\sigma_D}{\gamma_{Mf}} = \frac{41}{1.35} = 30.4 \, MPa$$

$$N = 10 \cdot 365 \cdot 15 = 54750$$

$$N = 5 \cdot 10^6 \cdot \left[\frac{30.4}{0.967 \cdot P}\right]^3 \qquad \Rightarrow P \leq 141.6 \, kN$$

Total weight per lorry:

$$\frac{3 \cdot 141.6}{9.81} = 43.3 \, ton$$

Literature

[1] Åkesson, B.: Forsmobron – en undersökning rörande bärförmåga, kondition och livslängd. (The Forsmo Bridge – an investigation concerning load-carrying capacity, condition and life-span expectancy. In Swedish). Chalmers University of Technology, Department of Structural Engineering, Division of Steel and Timber Structures, Publ. S 90:2, Göteborg 1990.

[2] Åkesson, B.: Gideälvsbron – statisk bärförmåga samt utmattningshållfasthet. Delrapport 1: Underlag till mätning och undersökning på plats. (The Gide Älv Bridge – static load-carrying capacity and fatigue strength. Report 1: Preparation for field testing and investigation. In Swedish). Chalmers University of Technology, Department of Structural Engineering, Division of Steel and Timber Structures, Int. skr. S 90:9, Göteborg 1990.

[3] Åkesson, B.: Gideälvsbron – statisk bärförmåga samt utmattningshållfasthet. Delrapport 2: Resultat av mätning och undersökning på plats. (The Gide Älv Bridge – static load-carrying capacity and fatigue strength. Report 2: The results from field testing and investigation. In Swedish). Chalmers University of Technology, Department of Structural Engineering, Division of Steel and Timber Structures, Publ. S 91:1, Göteborg 1991.

[4] Åkesson, B.: Malmbanan – teoretisk analys av uppkomna utmattningssprickor på fem stycken stålbroar. (The "Ore line" – a theoretical analysis of fatigue cracking in five railway bridges. Report 1: Preparation for field testing and investigation. In Swedish). Chalmers University of Technology, Department of Structural Engineering, Division of Steel and Timber Structures, Int. skr. S 91:5, Göteborg 1991.

[5] Åkesson, B.: Undersökning av järnvägsbroarna vid Laholm over Lagan och Laholms kvarnränna – bärförmåga och sprickkontroll. (An investigation of the railway bridges at Laholm over Lagan and The Laholm Mill Canal – load- carrying capacity and crack control. In Swedish). Chalmers University of Technology, Department of Structural Engineering, Division of Steel and Timber Structures, Rapport, Göteborg 1989.

[6] Lundgren, S. – Olsson, E.: SUPERINF – datorstöd vid analys och dimensionering av broar med avseende på utmattning. (SUPERINF – a computer programme support when analyzing and designing bridges against fatigue. In Swedish). Chalmers University of Technology, Department of Structural Engineering, Division of Steel and Timber Structures, Int. skr. S 90:5, Göteborg 1990.

156 Literature

[7] Åkesson, B.: Äldre järnvägsbroar i stål – bärförmåga, condition och livslängd. (Older railway bridges in steel – load-carrying capacity, condition and life-span expectancy. In Swedish). Chalmers University of Technology, Department of Structural Engineering, Division of Steel and Timber Structures, Int. skr. S 91:2, Göteborg 1991.

[8] Shen, Dafeng – Åkesson, B.: Maunojokk – a FEM-analysis of the out-of-plane deformations of a steel railway bridge. Chalmers University of Technology, Department of Structural Engineering, Division of Steel and Timber Structures, Int. skr. S 92:1, Göteborg 1992.

[9] Åkesson, B.: Hällnäs – Storuman. Bärförmåga, condition och livslängd hos 21 järnvägsbroar. (Hällnäs – Storuman. Load-carrying capacity, condition and life- span expectancy of 21 railway bridges. In Swedish). Chalmers University of Technology, Department of Structural Engineering, Division of Steel and Timber Structures, Int. skr. S 92:5, Göteborg 1992.

[10] Åkesson, B.: Fullskaleförsök avseende utmattning hos nitade balkar från järnvägsbroar – en litteraturstudie. (Full-scale fatigue testing of riveted railway bridge girders – a literature survey. In Swedish). Chalmers University of Technology, Department of Structural Engineering, Division of Steel and Timber Structures, Int. skr. S 93:4, Göteborg 1993.

[11] Fisher, J.W. – Struik, J.H.A.: Guide to design criteria for bolted and riveted joints. John Wiley & Sons, New York 1974.

[12] Forrest, P.G.: Fatigue of metals. Pergamon Press Ltd, London 1962.

[13] Kennedy, A.J.: Processes of creep and fatigue in metals. Oliver and Boyd Ltd, Edinburgh 1962.

[14] Almen, J.O. – Black, P.H.: Residual stresses and fatigue in metals. McGraw-Hill Book Company, Inc., New York 1963.

[15] Fatigue design handbook. Society of Automative Engineers, Inc., Warrendale, Pennsylvania 1988.

[16] Suresh, S.: Fatigue of materials. Cambridge University Press, Cambridge 1991.

[17] Fairbairn, W.: Experiments to determine the effect of impact, vibratory action, and long-continued changes of load on wrought-iron girders. Philosophical Transactions of the Royal Society, London, 154, 1864, pp 311–325.

[18] Reemsnyder, H.S.: Fatigue life extension of riveted connections. Journal of the Structural Division, Proceedings of the American Society of Civil Engineers, Vol. 101, No. ST12, 1975, pp 2591–2608.

[19] Wyly, L.T. – Scott, M.B.: An investigation of fatigue failures in structural members of ore bridges under service loading. Proceedings of the American Railway Engineering Association, Vol. 57, 1956, pp 175–297.

[20] Rabemanantsoa, H. – Hirt, M.A.: Comportement à la fatigue de profilés laminés avec semelles de renfort rivetées. ICOM Report 133, EPFL (Swiss Federal Institute of Technology), Lausanne 1984.

[21] Out, J.M.M. – Fisher, J.W. – Yen, B.T.: Fatigue strength of weathered and deteriorated riveted members. Transportation Research Record 950, Vol. 2, 1984, pp 10–20.

[22] Baker, K.A. – Kulak, G.L.: Fatigue of riveted connections. Canadian Journal of Civil Engineering, Vol. 12, 1985, pp 184–191.

[23] Baker, K.A. – Kulak, G.L.: Fatigue strength of two steel details. The University of Alberta, Edmonton, Department of Civil Engineering, Structural Engineering Report No. 195, 1982.

[24] Fisher, J.W. – Yen, B.T. – Wang, D. – Mann, J.E.: Fatigue and fracture evaluation for rating riveted bridges. NCHRP No. 302, Transportation Research Board, Washington, 1987.

[25] Abe, H.: Fatigue strength of corroded steel plates from old railway bridge. IABSE Symposium, Lisbon, 1989, pp 205–210.

[26] Fisher, J.W. – Yen, B.T. – Wang, D.: Fatigue strength of riveted bridge members. Journal of Structural Engineering, Vol. 116, No. 11, 1990, pp 2968–2981.

[27] Brühwiler, E. – Smith, I.F.C. – Hirt, M.A.: Fatigue and fracture of riveted bridge members. Journal of Structural Engineering, Vol. 116, No. 1, 1990, pp 198–214.

[28] Brühwiler, E. – Hirt, M.A.: Das ermüdungsverhalten genieteter brückenbauteile. Stahlbau 1/1987, pp 1–8.

[29] Mang, F. – Bucak, Ö.: Experimental and theoretical investigations of existing railway bridges. IABSE Workshop, Lausanne, 1990, pp 59–70.

[30] Mang, F. – Bucak, Ö.: Untersuchung an bestehenden brückenbau-werken im sinne von E-DS 805, sowie spezielle institutsversuche an einem originaltragwerk (anno 1877) und an seinen bauteilen. Der Metallbau im Konstruktiven-Ingenieurbau (K-I), Festschrift Prof. Tekn. Dr. Rolf Baehre zum sechzigsten geburtstag, Karlsruhe, 1988, pp 545–578.

[31] Mang, F. – Bucak, Ö.: Remaining fatigue life of old steel bridges – theoretical and experimental investigations of railway bridges. International Symposium on Fatigue and Fracture in Steel and Concrete Structures, Madras, 1991, pp 971–991.

[32] Mang, F. – Bucak, Ö.: Application of the S-N line concept for the assessment of the remaining fatigue life of old bridge structures. Second International Conference on Bridge Management, Guildford, 1993, pp 821–832.

[33] Pietraszek, T. – Oommen, G.: Static and dynamic behaviours of an 85-year old steel railway bridge. Canadian Journal of Civil Engineering, Vol. 18, No. 2 (1991), pp 201–213.

[34] Andersson, H.: Utmattning – teori, provning och dimensionering. Statens Provningsanstalt, Borås 1989.

[35] Engesvik, K.: Assessment of conditions and future service life of a railway bridge. IABSE Congress Helsinki, 1988, pp 373–378.

[36] Eriksson, K.: Toughness requirements for older structural steels. IABSE Workshop Lausanne, 1990, pp 95–101.

[37] Barsom, J.M. – Rolfe, S.T.: Fracture & fatigue control in structures. Prentice-Hall, Inc., Englewood Cliffs, New Jersey, 1987.

[38] Gurney, T. – Maddox, S.: An alternative to Miner's rule for cumulative damage calculations? IABSE Workshop Lausanne, 1990, pp 189–198.

[39] Maarschalkerwaart, H.M.C.M.: Determination of inspection intervals for riveted structures. IABSE Workshop Lausanne, 1990, pp 335–344.

[40] McGuire, W.: Steel structures. Prentice-Hall, Inc., Englewod Cliffs, New Jersey 1968.

[41] Rolfe, S. – Sorem, W. – Wellman, G.W.: Fracture control in the transition-temperature region of structural steels. Journal of Constructional Steel Research 12 (1989), pp 171–195.

[42] Yamada, K. – Miki, C.: Recent research on fatigue of bridge structures in Japan. Journal of Constructional Steel Research 13 (1989), pp 211–222.

[43] Berger, R.H. – Byrd – Tallamy – McDonald – Gordon, L.T.: Extending the service life of existing bridges. Transportation Research Record 664 (1978), pp 47–55.

[44] Hirt, M.: Fatigue considerations for the design of railroad bridges. Transportation Research Record 664 (1978), pp 86–92.

[45] Fisher, J.W. – Pense, A.W. – Slockbower, R.E. Hausmann, H.: Retrofitting fatigue damaged bridges. Transportation Research Record 664 (1978), pp 102–109.

[46] Grattesat, G.: Rating and evaluation of remaining life of bridges. IABSE Symposium Washington DC, 1982, pp 75–89.

[47] Siebke, H.: Supervision and inspection of the structures of the German Federal Railway. IABSE Symposium Washington DC, pp 3–13.

[48] Seong, C.K.: Fatigue resistance of riveted steel truss bridge members and joints. Lehigh University, Bethlehem, Pennsylvania 1983.

[49] Wiriyachai, A.: Impact and fatigue in open deck railway truss bridges. Illinois Institute of Technology, Chicago 1980.

[50] Karren, K. – Mahendrasinhji, G.: Strain hardening and ageing in cold-formed steel. Journal of the Structural Division, Vol. 101, No. ST1 (1975), pp 187–200.

[51] Verma, K. – McNamara, M.: A conceptual approach to prevent crack-related failure of steel bridges. Engineering Journal AISC, First quarter (1988).

[52] Fazio, R. – Fazio, A.: Rivet replacement criteria. Transportation Research Record 950, Vol. 1 (1984), pp 176–182.

[53] Hirt, M. – Yamade, K.: Fatigue life estimation using fracture mechanics. IABSE Colloquium, Lausanne 1982, pp 361–368.

[54] Yongji, S. – Yenman, Y. – Zegan, C.: Fatigue failures of steel railway bridges in China. IABSE Colloquium, Lausanne 1982, pp 516–524.

[55] van Maarschalkerwaart, H.: Fatigue behaviour of riveted joints. IABSE Colloquium, Lausanne 1982, pp 691–698.

[56] Smith, I. – Hirt, M.: Fatigue-resistant steel bridges. Journal of Constructional Steel Research 12 (1989), pp 197–214.

[57] Drew, F.: Recorded stress histories in railroad bridges. Journal of the Structural Division, Vol. 94, No. ST12 (1968), pp 2713–2724.

[58] Piraprez, E.: The effect of prying stress ranges on fatigue behaviour of bolted connections: The state-of-the-art. Journal of Constructional Steel Research, 27 (1993), pp 55–68.

[59] Carter, J.W. – Lenzen, K.H. – Wyly, L.T.: Fatigue in riveted and bolted single-lap joints. Transactions of the ASCE, Vol. 120 (1955), pp 1353–1388.

[60] Fernlund, I.: A method to calculate the pressure between bolted or riveted plates. Transactions of Chalmers University of Technology, No. 245, Gothenburg 1961.

[61] Mang, F. – Bucak, Ö.: Damage to old bridge structures caused by collision by vehicles. Bridge management 2. Thomas Telford, London 1993, pp 951–963.

[62] Herzog, M.: Abschätzung der restlebesdauer älterer genieteter eisenbahnbrücken. Stahlbau 10/1985, pp 309–312.

[63] Sugawara, N. – Oshima, T. – Mikami, S. – Sugiura, S.: On the accuracy improvement in ultrasonic inspection by using computer graphics and waveform analysis. Structural Engineering (Earthquake Engineering, Vol. 9. No. 4, 1993, pp 247–256.

[64] Riveting experience – In a throwback to an earlier age, a small bridge in Winchester, MA, is being fastened with rivets. Modern Steel Construction, September 1993, pp 34–37.

[65] Zuraski, P.D.: Service performance of steel bridges compared to fatigue-life predictions. Journal of Structural Engineering, Vol. 119, No. 10, 1993, pp 3056–3068.

[66] Fisher, J.W. – Menzemer, C.: Bridge repair methods – US/Canadian practice. NATO Advanced Research Workshop on Bridge Evaluation, Repair and Rehabilitation (Baltimore 1990). NATO ASI Series E: Applied Sciences – Vol. 187, pp 495–512.

[67] Fisher, J.W. – Daniels, J.H.: An investigation of the estimated fatigue damage in members of the 380-ft main span, Fraser River Bridge. American Railroad Engineering Association (AREA), Proceedings, 75th Technical Conference (Chicago 1976), Vol. 77, pp 577–597.

[68] Prine, D.W. – Hopwwod, T.: Improved structural monitoring with acoustic emission pattern recognition. Bridge evaluation, repair and rehabilitation. Kluwer Academic Publishers, Netherlands 1990, pp 187–199.

[69] Hahin, C. – South, J.M. – Mohammadi, J. – Polepeddi, R.K.: Accurate and rapid determination of fatigue damage in steel bridges. Journal of Structural Engineering, Vol. 119, No. 1, 1993, pp 150–168.

[70] Wagh, V.P. – Abrahams, M.J.: Fatigue evaluation of a riveted railroad bridge. Proceedings of the Sessions Related to Structural Materials – Structures Congress (San Francisco 1989), pp 179–188.

[71] Smith, R.A.: The Versailles railway accident of 1842 and the first research into metal fatigue. Fatigue 90. Materials and Component Engineering Publications, Birmingham 1990, pp 2033–2041.

[72] Nagase, Y. – Suzuki, S.: On the decrease of fatigue due to small prestrain. Journal of Engineering Materials and Technology, Vol. 114, 1992, pp 317–322.

[73] Ananthanarayana, N. – Jain, B.K.: Fatigue life of early steel girder bridges on Indian railways. Symposium on Fatigue and Fracture in Steel and Concrete Structures, Madras 1991, pp 1015–1033.

[74] Brandes, K.: Crack growth tetsts to assess the remaining fatigue life of old steel bridges. IABSE Workshop Remaining Fatigue Life of Steel Structures, Lausanne 1990, pp 103–109.

[75] Kayser, J.R. – Nowak, A.S.: Capacity loss due to corrosion in steel-girder bridges. Journal of Structural Engineering, Vol. 115, No. 6, 1989, pp 1525–1537.

[76] Barsom, J.M.: Fatigue-crack propagation in steels of various yield strengths. Journal of Engineering for Industry, November 1971, pp 1190–1196.

[77] Schilling, C.G. – Klippstein, K.H. – Barsom, J.M. – Novak, S.R. – Blake, G.T.: Low-temperature tests of simulated bridge members. Journal of the Structural Division, Vol. 101, No. ST1, 1975, pp 31–48.

[78] Brühwiler, E. – Hirt, M.A. – Morf, U. – Huwiler, R.: Bewertung der spontanbruchgefahr angerissener brückenbauteile aus schweisseisen. Stahlbau 58 (1989), H. 1, pp 9–16.

[79] Chajes, A. – Britvec, S.J. – Winter, G.: Effects of cold-straining on structural sheet steels. Journal of the Structural Division, Vol. 89, No. ST2, 1963, pp 1–32.

[80] Wasén, J. – Karlsson, B.: Influence of prestrain and ageing on near-threshold fatigue crack growth in fine-grained dual-phase steels. International Journal of Fatigue, Vol. 11, No. 6, 1989, pp 395–405.

[81] Tanner, P. – Hirt, M.A.: Überlegungen zur restlebensdauer schweisseiserner brücken am beispiel der Basler Wettsteinbrücke. Stahlbau 60 (1991), H. 7, pp 211–219.

[82] Nishimura, A. – Tajima. J. – Yamasaki, T. – Kikukawa, S.: Ageing of high-strength bolted joints in long service. IABSE Symposium Washington DC, 1982, pp 149–154.

[83] Gul, Y.P. – Shukis, I.Z.: Effect of strain ageing on the temperature dependence of the components of impact toughness. Metal Science and Heat Treatment, Vol. 17, No. 7–8, 1975, pp 715–716.

[84] Gul, Y.P.: Effect of carbon and nitrogen on hardening and embrittlement of low- carbon steel during ageing. Metal Science and Heat Treatment, Vol. 17, No. 7–8, 1975, pp 553–557.

[85] Brandes, K.: Untersuchungen zum rissfortschritt an einer stahlbrücke. IABSE Congress Helsinki, 1988, pp 361–366.

[86] Grundy, P. – Chitty, G.B.: Remaining life of a suite of railway bridges. IABSE Workshop Lausanne, 1990, pp 45–56.

[87] Schindler, H-J. – Morf, U.: Toughness and fracture behaviour of obsolete wrought bridge steel. IABSE Workshop Lausanne, 1990, pp 85–93.

[88] Cullimore, M.S.G.: Fatigue of HSFG bolted joints – effects of design parameters. IABSE Workshop Lausanne, 1990, pp 715–723.

[89] Wolchuk, R. – Mayrbaurl, R.M.: Stress cycles for fatigue design of railroad bridges. Journal of the Structural Division, Vol. 102, No. ST1, 1976, pp 203–213.

[90] Wirsching, P.H. – Light, M.C.: Fatigue under wide band random stresses. Journal of the Structural Division, Vol. 106, No. ST7, pp 1593–1607.

[91] Zwerneman, F.J. – Frank, K.H.: Fatigue damage under variable amplitude loads. Journal of Structural Engineering, Vol. 114, No. 1, pp 67–83.

[92] Zimmer, R. – Katzschner,E.: Einige bemerkungen zu älterer baustählen im eisenbahnbrückenbau. Signal und Schiene, 1/1981, pp 38–43.

[93] Tutzschky, G.: Versuche zur bestimmung der materialeigenschaften an alten brückenstaählen. Signal und Schiene, 5/1967, p 187.

[94] Kunz, P. – Hirt, M.A.: Optimisation of inspection intervals with respect to remaining fatigue life. Second International Conference on Bridge Management, Surrey, 1993, pp 56–65.

[95] Sedlacek, G. – Hensen, W.: Nouvelles méthodes de calcul pour la réhabilitation des ponts métalliques anciens. Construction Métallique, No. 3, 1992, pp 3–12.

[96] Torng, T.Y. – Wirsching, P.H.: Fatigue and fracture reliability and maintainability process. Journal of Structural Engineering, Vol. 117, No. 12, 1991, pp 3804–3822.

[97] Verma, K.K. – Beckmann, F.R.: High-strength bolts for bridges. Engineering Journal (AISC), First Quarter, 1992, pp 4–11.

[98] Eriksson, K.: Fatigue assessment of service damaged and corroded members of structural steel. VTT Symposium 130 Fatigue Design, Helsinki 1992, pp 215–227.

[99] Sedlacek, G. – Hensen, W. – Bild, J. – Dahl, W. – Langenberg, P.: Verfahren zur ermittlung der sicherheit von alten stahlbrücken unter verwendung neuester erkenntnisse det werkstofftechnik. Bauingenieur 67 (1992), pp 129–136.

[100] Yamasaki, T. – Kawai, Y. – Maeda, Y.: Fatigue life of welded and bolted repair parts. Journal of Structural Engineering, Vol. 110, No. 10, 1984, pp 2499–2512.

[101] Wilson, W.M. – Thomas, F.P.: Fatigue tests of riveted joints. University of Illinois, Engineering Experiment Station, Bulletin No. 302, Urbana 1938.

[102] Wilson, W.M. – Coombe, J.V.: Fatigue tests of connection angles. University of Illinois, Enginnering Experiment Station, Bulletin No. 317, Urbana 1939.

[103] Wilson, W.M. – Munse, W.H. – Cayci, M.A.: A study of the practical efficiency under static loading of riveted joints connecting plates. University of Illinois, Engineering Experiment Station, Bulletin No. 402, Urbana 1948.

[104] Schilling, C.G. – Klippstein, K.H.: New method for fatigue design of bridges. Journal of the Structural Division, Vol. 104, No. ST3, 1978, pp 425–438.

[105] Case histories involving fatigue and fracture mechanics. ASTM Special Technical Publications 918, Philadelphia 1985.

[106] Fatigue at low temperatures. ASTM Special Technical Publication 857, Philadelphia 1983.

[107] Yen, B.T. – Huang, T. – Lai, L-Y. – Fisher, J.W.: Manual for inspecting bridges for fatigue damage conditions. Lehigh University, Department of Civil Engineering, Fritz Engineering Laboratory, report No. 511-1, Bethlehem 1990.

[108] Fisher, J.W.: Fatigue and fracture in steel bridges – case studies. John Wiley & Sons, New York 1984.

[109] Wåle, J. – Bergh, S.: Inverkan av kalldeformation på segheten hos stål. (The influence of cold-straining on the toughness of steel. In Swedish). AB Statens Anläggningsprovning, SA/FoU-rapport 90/07, Stockholm 1990.

[110] Parola, J.F. – Chesson Jr, E. – Munse, W.H.: Effect of bearing pressure on fatigue strength of riveted connections. University of Illinois, Engineering Experiment Station, Bulletin No. 481, Urbana 1965.

[111]   Wilson, W.M. – Bruckner, W.H. – McCrackin Jr, T.H.: Tests of riveted and welded joints in low-alloy structural steels. University of Illinois, Engineering Experiment Station, Bulletin No. 337, Urbana 1942.

[112]   Forsberg, F.: Utmattningshållfasthet hos äldre konstruktionsstål med korrosionsskador. (Fatigue strength of older structural steels with corrosion damage. In Swedish). Tekniska Högskolan i Luleå, Avdelningen för Stålbyggnad, Examensarbete 1993:064 E, Luleå 1993.

[113]   Yin, W.S. – Fang, Q.H. – Wang, S.X. – Wang, X.H.: Fatigue strength of high- strength bolted joints. IABSE Colloquium, Lausanne 1982, pp 707–714.

[114]   Bestämmelser rörande tillverkning, leverans och uppsättning under åren 1910–1912 af järnöfverbyggnad för ny bro öfver Ångermanälfven vid Forsmo. (Rules and regulations regarding manufacturing, deliverance and erection during the years 1910-1912 of the iron superstructure of the new bridge over the river Ångermanälven at the village Forsmo. In Swedish). Kungliga Järnvägsstyrelsen, Bankonstruktionsbyrån, Stockholm ~1910.

[115]   Beskrifning öfver förstärkning af bron öfver Indalsälfven vid Ragunda åren 1910-1912. (A description of the reinforcement of the bridge over the river Indalsälven at the village Ragunda during the years 1910-1912. In Swedish). Kungliga Järnvägsstyrelsen, Bankonstruktionsbyrån, Stockholm 1915.

[116]   Banlära – Järnvägars byggnad och underhåll. (Railway engineering – Construction and maintenance of railways. In Swedish). Kungliga Järnvägsstyrelsen, Stockholm 1916.

[117]   Normalbestämmelser för järnkonstruktioner till byggnadsverk. (Rules and regulations regarding iron structures in building construction. In Swedish). Statens offentliga utredningar 1938:37, Kommunikationsdepartementet, Stockholm 1938.

[118]   Kim, J.B. – Brungraber, R.H. – Kim, R.H.: Recycling bridges. Civil Engineering, November 1988, pp 58–59.

1995–2009

[119]   Åkesson, B. – Edlund, B.: Extended service life for riveted railway bridges? Nordic Steel Construction Conference, Malmö 1995.

[120]   Åkesson, B. – Edlund, B.: Fatigue life of riveted railway bridges. IABSE Symposium, San Francisco 1995.

[121]   Persson, D.: Ståltyper i äldre svenska järnvägsbroar. (Steel types in old railway bridges in Sweden. In Swedish). Chalmers University of Technology, Department of Structural Engineering, Division of Steel and Timber Structures, Int. skr. S 95:7, Göteborg 1995.

[122]   Adamson, D.E. – Kulak, G.L.: Fatigue tests of riveted bridge girders. Structural Engineering Report No. 210. Department of Structural Engineering, University of Alberta, Edmonton , 1995.

[123]   Dibattista, J.D. – Kulak, G.L.: Fatigue of riveted tension members. Structural Engineering Report No. 211. Department of Structural Engineering, University of Alberta, Edmonton 1995.

[124] Zhou, Y.E. – Fisher, J.W. – Sweeney, R.A.P.: Examination of fatigue strength (S_r – N) curves for riveted bridge members. 12th International Bridge Conference, Engineering Society of Western Pennsylvania, Pittsburgh, June 1995.

[125] Åkesson, B. – Edlund, B.: Remaining fatigue life of riveted railway bridges. Stahlbau 65, Heft 11, 1996.

[126] Shafie, M.A. – Sabardin, L.N.: Ageing and brittle fracture in steel. Chalmers University of Technology, Department of Structural Engineering, Division of Steel and Timber Structures, Int. skr. S 97:8, Göteborg 1997.

[127] Helmerich, R. – Brandes, K. – Herter, J.: Full scale laboratory fatigue tests on riveted railway bridges. Evaluation of existing steel and composite bridges, IABSE Workshop, Lausanne 1997.

[128] Kadir, Z.A. Riveted joints – full-scale fatigue tests. Chalmers University of Technology, Department of Structural Engineering, Division of Steel and Timber Structures, Int. skr. S 97:7, Göteborg 1997.

[129] Al-Emrani, M. – Crocetti, R. – Åkesson, B. – Edlund, B.: Fatigue crack arrest in riveted bridge girders using stop-holes. Eurosteel, Prague 1999.

[130] Crocetti, R. – Al-Emrani, M. – Åkesson, B. – Edlund, B.: Constant amplitude fatigue limit for riveted girders. Eurosteel, Prague 1999.

[131] Paasch, R.K. – DePiero, A.H.: Fatigue crack modelling in bridge deck connection details. Final Report, SPR 380, December 1999.

[132] Crocetti, R.: On some fatigue problems related to steel bridges. Chalmers University of Technology, Department of Structural Engineering, Göteborg 2001.

[133] Xie, M. – Bessant, G.T. – Chapman, J.C. – Hobbs, R.E.: Fatigue of riveted bridge girders. The Structural Engineer, 79 (9), 2001.

[134] Al-Emrani, M.: Fatigue in riveted railway bridges – A study of the fatigue performance of riveted stringers and stringer-to-floor-beam connections. Chalmers University of Technology, Department of Structural Engineering, Göteborg 2002.

[135] Al-Emrani, M. – Åkesson, B.: Experimentelle untersuchung des ermüdungsverhaltens genieteter träger. Stahlbau 71, Heft 2, 2002.

[136] Imam, B.: Fatigue assessment of riveted railway bridges – literature review. University of Surrey, School of Engineering, July 2003.

[137] Al-Emrani, M. – Åkesson, B. – Kliger, R.: Examination of some overlooked secondary effects in open-deck truss bridges. Structural Engineering International, Vol. 14, No. 4, November 2004.

[138] Guidance on fatigue assessment of steel structures using EN1993-1-9. A guideline for fatigue design. ECCS – Technical Committee 6 – Fatigue, First Edition, May 2005, N° XXX (Draft).

[139] Matar, E.B – Greiner, R.: Fatigue tests for a riveted steel railway bridge in Salzburg. Structural Engineering International, August 2006.

[140] Matar, E.B.: Evaluation of fatigue category of riveted steel bridge connections. Structural Engineering International 1/2007.

[141] Imam, B.M. – Righiniotis, T.D. – Chryssanthopoulos, M.K.: Probabilistic fatigue evaluation of riveted railway bridges. Journal of Bridge Engineering, ASCE, May/June 2008.

[142] Righiniotis, T.D. – Imam, B.M. – Chryssanthopoulos, M.K.: Fatigue analysis of riveted railway bridge connections using the theory of critical distances. Engineering Structures 30 (2008).

[143] Siriwardane, S. – Ohga, M. – Dissanayake, R. – Taniwaki, K.: Application on new damage indicator – based sequential law for remaining fatigue life estimation of railway bridges. Journal of Constructional Steel Research 64 (2008).

[144] Kühn, B. – Lukic', M. – Nussbaumer, A. – Günther, H-P. – Helmerich, R. – Herion, S. – Kolstein, M.H. – Walbridge, S. – Androic, B. – Dijkstra, O. – Bucak, Ö.: Assessment of existing steel structures: Recommendations for estimation of remaining fatigue life. Background documents in support to the implementation, harmonization and further development of the Eurocodes. Joint Report. Prepared under the J.R.C. – ECCS cooperation for the evolution of Eurocode 3 (Programme of CEN / TC 250). Editors: G. Sedlacék, F. Bijlaard, M. Géradin, A. Pinto, S. Dimova. First edition, February 2008, EUR 23252 EN- 2008.

[145] Larsson, T.: Fatigue assessment of riveted bridges. Doctoral thesis, Luleå University of Technology, Department of Civil and Environmental Engineering, Division of Structural Engineering, Luleå, February 2009.

[146] Pipinato, A. – Pellegrino, C. – Bursi, O.S. – Modena, C.: High-cycle fatigue behavior of riveted connections for railway metal bridges. Journal of Constructional Steel Research 65 (2009).

[147] De Jesus, A.M.P. – Pinto, H. – Fernández-Canteli, A. – Castillo, E. – Correia, J.A.F.O.: Fatigue assessment of a riveted shear splice based on a probabilistic model. International Journal of Fatigue (accepted manuscript in September 2009).

[148] Siriwardane, S. – Ohga, M. – Kaita, T. – Dissanayake, R.: Grain-scale plasticity based model to estimate fatigue life of bridge connections. Journal of Constructional Steel Research 65 (2009).

[149] Caglayan, B.O. – Ozakgul, K. – Tezer, O.: Fatigue life evaluation of a through-girder steel railway bridge. Engineering Failure Analysis 16 (2009).

Index

AASHTO 12, 15–16, 114, 143–144
Abe, H. 8, 17–18, 21, 136
acoustic emission 70–71
Adamson, D.E. 143
ageing 2, 4–5, 58, 64, **65**, 71, 74, 87, 89–90, 123–127, 136–138, 140
Al-Emrani, M. 144
allowable stresses 1, 10, 91
aluminium 68
Ångermanälven 73–74
anisotropy 47
arch reinforcement method 72
AREA 12
axle spacing 76, 93
axle weight 1, 91–98, 130, 149

Baker, K.A. 8, 14–16, 20, 134
ballast 2, 72, 75, 99
bearing ratio 7, 39
bearing strength 18
bearing stress 14, 39
bearings 46, 79–81, 86–88
bending stiffness 15, 56–57, 81, 106, 123
Bessemer process 67
Bethlehem 3
Blumberg Bridge 19
bogie-axle 109, 130–133
bolts 10–12, 14–15, 20–21, 23, 25, 32, 54, 72, 136, 141, 144
bracings 109
braking stiffeners 46, 103
Brandes, K. 143
Bressant, G.T. 144
British Standard 62
brittle fracture XI, XII, 2–3, 5, 17, 21, 58, 64–65, 67, 70, 72, 74–75, 100–101, 134, 139
Brühwiler, E. 8, 17–18, 21

Bucak, Ö. 8, 18–19
Bursi, O.S. 145

carbon 27, 31–32, 47, 58–59, 63, 68, 91, 127, 138
Carter, J.W. 42, 50
category C 12–20, 114, 122, 134–135, 137, 143–145
category D 12–20, 114, 122, 134–135, 137, 143–145
centrifugal force 57
Chapman, J.C. 144
Charpy V-notch tests 61–63, 67, 90, 100, 126, 137–138
chlorides 69
chromium 127
circumferential ("hoop") stresses 25–26, 41–42
clamping force XI, 5–7, 10, 12, 14 20, 23, **25**
clamping ratio 26, 129–130, 138
cleavage fracture 58, 61, 65
coating 7
cobalt 127
cohesive powers 9
cold-work damage 39, 123, 126
cold-working 65–67
composite action 5, 8, 13, 21, 52, 62, 70, 104, 112, 115, 136–137, 139
concrete cladding 72
congregation points 65
constant-moment region 52, 110–112, 115
copper 127
corrosion 2–3, 5, 14, 17, 19, 21, 25, 42, **68**, 70–72, 74–75, 108, 123, 134, 138
cover plates 12–13, 143–144
crack growth 14, 21, 59–60, 64, 100, 123, 134

crack initiation 14, 39, 60, 109, 114–116, 119, 123, 137, 145
crack propagation rate 4, 14, 21, 59–60, 65, 67, 70–71, 101, 112, 115–116, 119–120, 123, 137, 140, 144, 145
creep 30, 43
crevice corrosion 69, 75
critical strain rate 64
Crocetti, R. 144
crookedness 45
crystalline structure 9, 123

defect fatigue resistance 136
deflection 45–46, 56–57, 71, 74–75, **75**, 79–80, 82, 85, 109–110, 119, 123–124, 138
deformability 30
design specifications 8, 30
Dibattista, J.D. 143
diffusion 65, 89
dislocation 39, 43, 58, 65, 68, 75
displacement gauges 79
driving temperature 29, 32
ductility XI, 30–31, 39, 47, 58, 60, 63–67, 139
dynamic amplification (factor) 2, 3, 46, **56**, 74–76, **79**, 91, 99, 138, 139

ECCS 12, 114, 144
effective deformation length 128
eigenfrequency 57, 81–82
electric pressure riveter 25
elongation 7, 27, 29–31, 40, 46, 50, 54, 87–90, 125–127, 129, 138
embrittlement 2, 39, 67, 72
end fastener 7
EPFL 3
equivalent bogie-axle passages 131, 133
equivalent freight train passages 94, 98, 130, 140
Eriksson, K. 62

Fairbairn, W. 8–10, 19
fatigue crack control 3
fatigue cracking XI, 2, 6, 10, 24, 45, 47–51, 56, 70, 72, 86, 99, 110, 123, 140–141, 143–144
fatigue damage accumulation (factor) XI, 3–5, 20–21, 49, 51, 58, 74, 77, 91–92, 94, 97–99, 102–103, 122, 130, 132, 134–136, 138–140, 144, 146, 148–149

fatigue fracture value 19, 60, 103, 119, 122
fatigue limit 3–4, 13, 17, 20–21, 48, 60–62, 67, 132, 139–140, 144, 148
Fernlund 25
Fisher, John W. 3, 8, 13–16, 21, 37–38, 61, 134–135, 143
flange-rivet connection 112–119, 136–137
flaw 39, 59
flexural rigidity 109
floor beam 5–6, 37, 49–56, 75, 83–85, 94, 96, 99, 103–104, 106, 136, 139, 144
floor-beam hangers 49, 52
forging 24
Forsmo Bridge 30, 73–74, 77–80, 84–86, 88, 90–94, 99
four-point bending 12–14, 17, 109, 143–144, 146
fracture lives 122
fracture mechanics 60, 62, 64, 91, **99**, 136, 138
fracture toughness XI, 21, 59–60, 62
fretting 14, 30, 42–43, 55, 70, 123, 130, 145
fretting corrosion 43
friction coefficient 32–33, 42
friction force 121
friction slip 55, 57, 71, 121, 139
frictional bond 14
friction-grip bolted joint 32–33, 54, 72, 136, 144

Gide Älv Bridge 73, 77, 79, 81–88, 90–91, 95–96, 99
grain size 30, 47, 58–59, 63, 87, 91, 138
grain structure 92
Greiner, R. 144
grip length 7, 26–27, 29, 31–32, 34, 46, 130
Hällnäs 73, 84, 87

hammer 24–25, 27, 30–31, 56, 66, 104, 108
hand hammers 24–25, 47, 56, 118
hand-hammered rivets 30, 47, 66, 104
HEB 1000 12
Helmerich, R. 143
Herter, J. 143
high-hydrogen electrodes 59
high-strength bolts 5, 10–12, 14–15, 20–21, 23, 32–33, 54, 72, 136, 141
highway bridges 5, 15, 17, 45, 47–48
Hirt, Manfred A. 3, 8, 12–13, 17–18
hit marks 2, 46, 62
Hobbs, R.E. 144

Holmsund 130
hoop rings 25–26
hoop stress 25–26, 42
horizontal bracing bar 17, 19, 103, 109, 136, 141
hydraulic jacks 109–110, 112–113, 127, 146
hydraulic (pressure) riveter 25, 27, 31, 47

impact 8, 9, 21, 56, 64, 126, 137–138
impact energy 61, 67, 90, 100, 126
impact toughness 61–62, 64, 126
impact transition temperature 63
impulse-echo method 84
impurities 47, 64, 68
inclination 36, 45
influence length XI, 5, 49, 56, 131, 140, 149–150
initial imperfections 45
inspection interval 71
inspection routines 4, 70–71, 102, 123, 141
iron-ore bridge 10–12

Kadir, Z.A. 144
Kaldo 68
killed steel 58, 68
Krautkrämer USK 7 84
Kulak, G.L. 8, 14–16, 20, 134, 143

Laholm Bridge 73, 84, 86
Lapland iron ore 97
large grains 58
lateral secondary bending 109
lateral stability 109
lattice bridge girder 57
lattice girders 17, 57, 143
Lausanne 3
LD 68
lead 25, 42, 69
Lehigh University 3
Lenzen, K.H. 42
Liberty ships 62
linseed oil (paint) 42, 69
load combinations 2
load-carrying capacity XI, 2–3, 5–6, 21, 25, 34, 36, 45, 47, 54, 67, 69–74, **91**
loading history XI, 5, 48–49, 92, 130–131, 135, 138
loading rate 58, 60, 64–65
loading rig 113
loose rivets 2, 4, 14, 46, 70, 108, 141

Losenhausenwerk 109
L-shaped profiles 9, 37

maintenance 2, 4, 69, **70**, 138, 141, 145
Mang, F. 8, 18–19
manganese 91, 127
Mann, J.E. 8, 15
Matar, E.B. 144
Martin 68, 127
Martin steel 127
mass inertia forces 56–57, 80, 82
material properties 46, 58, 65, 101
mean stress 17, 21, 135
mechanical treatment 30–32
metallurgical investigation 91
micro-cracks 39, 60
micro-defects 39, 63–65, 84, 100, 138
mild steel 17, 124, 137–138, 143
misalignment 34, 46, 75–76, 127
modulus of elasticity 29
Modena, C. 145
molybdenum 127
movable bridge bearings 46

National Rail Administration 1, 62, 92, 97
neck-rivet 110–112, 115–117, 119, 137
nickel 127
nitrogen 58–59, 65, 67–68, 74, 91, 127, 138
notch 15, 21, 39, 59–63, 65, 67, 69, 90, 126, 136–138, 149

open-deck construction 5
Out, J.M.M. 8, 13–14
Oxygen 47, 68–69

Pellegrino, C. 145
Perlite 58, 87, 91–92
phosphorus 47, 58, 68, 91, 127, 138
pin bolts 23
Pipinato, A. 145
pitch distance 25, 46
pitting corrosion 69, 75, 108
plastic flow 60–61, 65
plate thickness XI, 14, 24–27, 36–37, 43, 58–59, 62–63, 65, 100, 139
plate tolerance 32
pneumatic hammer 25, 27, 31, 47, 104
pre-heating 59
prestressing 35–36
prying 37–38, 43, 52–53, 55
puddle iron 23

Index

pulsating machine 109, 112, 146
punching 24, 39, 66

Rabemanantsoa, H. 8, 12–13, 17
radial stresses 25–26, 41–42
radius of curvature 57, 82
reaming 24
recovery length 128
red lead 25, 42, 69
redundancy XI, 5, 13–14, 17, 21, 60, 62, 100, 115, 120, 123, 137, 139
redundant fracture behaviour 45
Reemsnyder, H.S. 8, 10–12, 20, 134
relaxation 30, 42–43
residual clamping force 7, 32, 35, 129
residual stresses 63
resonant vibration 56–58, 75, 80
rimmed steel 47, 58, 68, 127
rivet installation 7
rivet pattern 7
rivet revision 47, 70
riveting machine 24–25, 104
road bridges 5, 15, 17, 45, 47–48
rust-protective paint 69

scale 25, 42
Scott, M.B. 10
secondary bending 43, 50, 89, 99, 109, 139
semikilled steel 58
semi-rigid behaviour 46, 54–55, 58
separation load 36
shear deformations 30, 42–43, 53, 55
shear fracture 58, 61,65
shear strength 7, **32**
shear stress range 17, 21, 52
shearing deformations 58
shop riveting 23, 25, 47, 55
silicon 68, 127
slag 47, 87, 91
slip bands 61, 63
slip planes 33–34
slip resistance 7
Smith, I.F.C. 8, 17–18
spectrum loading 10, 12, 136, 148
speed limits 94
Stahringen Bridge 19
standard deviation 20, 124–126, 129–130
steel quality 5, 46–47, 57, 65, 92, 138
steel type 5, 67–68, 74
stop hole 13, 136, 144
Storuman 84, 97

strain ductility 30, 39
strain gauges 75–76, 84
strain hardening 65, 126
stress concentration factor 38–41
stress intensity factor 64, 99–101
stress raiser 38, 52, 58–59, 66–67, 69, 110, 139
stress range 3–4, 8–10, 12–13, 15, 17, 20–21, 39, 43, 48–49, 51–52, 55–56, 58, 67, 75, 77, 95, 98, 109–111, 114–115, 117, 120–122, 131–132, 134–135, 139–140, 143–144, 146, 148–149
stress relaxation 43
stringer XI, 4–6, 13–14, 17–18, 37–38, 46, 49, 51–56, 72, 75, 79–80, 84–85, 94, 96, 103–104, 106–111, 113–128, 130–132, 134–140, 143–144
stringer-to-floor-beam connection 49, 53, 55, 144
stroke length 109–110, 112
structural damping 56
structural redundancy XI, 5, 17, 21, 60, 62, 100, 115, 120, 123, 137, 139
Struik, J.H.A. 37–38
struts 109
sub-punching 24
sulphur 47, 58, 91, 127
SUPERINF 76
superstructure 2, 19, 37, 66, 68, 72–73, 75, 86, 138
surface corrosion 69, 75, 108
surface finish 29, 33
surface treatment 29
Sweeney, R.A.P. 143

tangential velocity 57
temperature 4, 14, 17, 21, 29, 32, 43, 58–65, 67–69, 90, 100, 126–127, 134, 136–138, 140
tensile strength 30–32, 34, **35**, 58, 65–67, 87–88, 90, 126, 129, 138
tension test 92, **123**, 129, 137–138
Thomas process 67–68, 138
Thomas, F.P. 26–27, 30–31
titanium 127
track alignment 56, 75, 83
track irregularities 56, 75, 83
track superstructure 37
transition temperature 61–64, 67, 90, 126–127, 137–138

trough 72, 104
T-shaped profiles 37

UIC 71 78, 85, 92–96, 138
ultimate strength 19, 34
ultrasonic testing XI, 3, 70–71, 84, 86, 99, 136, 138, 140
Ume Älv (Åskilje) 73, 84, 88, 90–91, 99
Ume Älv (Lycksele) 73, 84, 88, 90–91, 99
utilization factor 91, 93–96

vanadium 127
Vänersborg 23
Vännäs 106–107, 110, 130
Vännäsby 53–54, 56, 103–107, 112, 131, 135, 137, 141, 144
vertical shear loading 130
vibratory action 8, 9
Vindelälven 53–54, 56, 73, 84, 88, 90–91, 99, 103–106, 112, 131, 135, 137, 141

Vindelälven (Åmsele) 73, 84, 88, 90–91, 99
visual inspection 70, **75**

Wang, D. 8, 15
wide-flange beams 14
Wilson, W.M. 26–27, 30–31
wind-bracing diagonals 136
Wöhler diagram 114, 134
work hardening 30
wrought iron 8–10, 17–18, 31, 143–144
Wyly, L.T. 10, 42

Xie, M. 144

Yen, B.T. 8, 13–15
yielding 34, 42–43, 58, 65–67
Yin, W.S. 40

Zhou, Y.E. 143